THE ART OF TIMING

By the same authors
MOON TIME

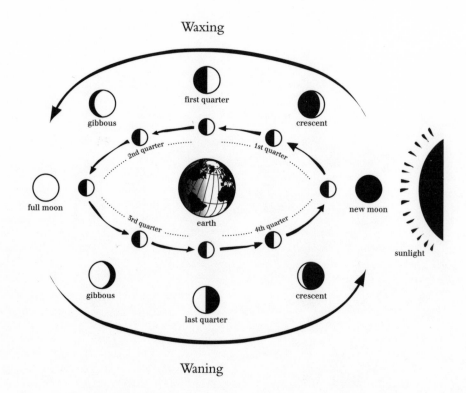

Waxing

first quarter

gibbous

crescent

2nd quarter

1st quarter

full moon

earth

new moon

3rd quarter

4th quarter

sunlight

gibbous

crescent

last quarter

Waning

The Waxing and Waning of the Moon.

Johanna Paungger & Thomas Poppe

THE ART OF TIMING
The Application of Lunar Cycles in Daily Life

Translated from the German by David Pendlebury

Index compiled by Ann Griffiths

SAFFRON WALDEN
THE C.W. DANIEL COMPANY LIMITED

Originally published by
Heinrich Hugendubel Verlag, Munich, Germany
under the title *Vom richtigen Zeitpunkt*

First published in Great Britain in 2000
by The C.W. Daniel Company Limited
1 Church Path, Saffron Walden,
Essex, CB10 1JP, United Kingdom

ISBN 0 85207 334 8

Illustrated and Designed by Jane Norman
Produced in association with Book Production Consultants plc, 25-27 High Street,
Chesterton, Cambridge, CB4 1ND
Typesetting by Cambridge Photosetting Services
Printed and bound by Hillman Press (Frome) Ltd, England

CONTENTS

The Moon as a Helper in the Home and Everyday Life

Foreword

For many years people have been asking me to pass on a body of knowledge with which I grew up from my earliest childhood – the knowledge of lunar rhythms, and the influences exerted on all forms of life on earth, which are indicated by the position and phases of the moon. I have to thank my grandfather for teaching me that instinct; perception and experience are the keys to many things in nature that science alone is unable to unveil.

I can still remember very well the first lecture I ever gave. People warned me that I would have to reckon with derision. Yet my inner conviction was so firm and I was also so sure of the support of my friends, that I didn't care how many of the audience laughed at me. Only one thing mattered to me. If even a single person was ready to receive this natural and self-evident material, then this would keep alive an ancient body of knowledge which has been maintained for centuries, by virtue of being handed down, tested and applied, and which especially at the present time could be of great value for us all and for the world in which we live.

The success of this first lecture encouraged me to give others; and now, after many such lectures, I am writing my experiences down. I am especially pleased at the openness which many people have shown towards the knowledge of lunar rhythms. If at first I saw many incredulous faces, there soon developed an explosive interest in the subject. Today many doctors and employers actually make use of the 'the art of good timing' in their professional work. For me there is absolutely nothing new about dealing with the phases of the moon; but I am happy that many people are once more placing their trust in this ancient knowledge.

I hope that you come with an open mind and are prepared to absorb this knowledge that can accompany you throughout the whole of your life, without your having to be forever looking up textbooks, guides and tables. A knowledge of lunar rhythms and their influence upon us should become second nature to you and give you a valuable gift which you can pass on to your children as a springboard for their own experience. Even if you did not grow up, as I did, with this understanding of the moon's rhythms, you now have the opportunity to accumulate your own experiences, and that is a thousand times more useful than examples in a book. After you have tried out some of the suggestions for living in harmony with the lunar calendar a few times, you will soon notice how much easier everything seems.

This book was written in collaboration with Thomas Poppe. I hope that it may make life easier for every person of good will who reads it, and that it may serve as a helpful companion in a great many of life's situations.

Johanna Paungger

Great adventures sometimes have trivial, unspectacular beginnings. For example, the ringing of a telephone.

"Hey, I've just met this woman that you might find really interesting. Maybe you could write a book together. She's called Johanna Paungger. I told her about you and she wants to meet you..."

"Write a book? What about?" I answered, somewhat irritably, as I was in the middle of grappling with a book I didn't like and my inspiration had deserted me.

"Wait and see, it's not so easy to explain."

That's all I needed.

Well, being inquisitive by nature, I agreed. I had no idea at the time what was heading my way.

Many events and experiences in my life deserve the description 'extraordinary, strange, uplifting, enriching', but my meeting with Frau Paungger does not fit into any of these categories. The quality of the encounter with her was so new to me that there were no standards available to me by which to measure it and place it in some corner or other of my thinking and feeling.

Not that anything unusual or sensational took place. We met in a forest café, spoke only sparingly about the subject of the proposed book, exchanged friendly remarks and anecdotes in order to surmount the initial distance between us, and philosophised about this and that. She said she had read one of my books and had gained the feeling that I was the right person to work with her on writing an account of an ancient body of knowledge. She told me about her Tyrolean homeland, her childhood as one of ten children in a family of mountain farmers, how she had moved to Munich. Again and again, almost incidentally, she slipped in allusions to a special knowledge, which was still widespread back home, and which her grandfather had communicated to her. The knowledge concerning the rhythms of the moon and their influence on nature, man, beast and plant.

One anecdote that casts a light on her apprenticeship with her grandfather remains in my memory. She related that the long years of learning had gone by almost without anything being said – only watching, observing, grasping, experiencing. But one day she had actually asked a question. I believe it was in connection with gathering a particular medicinal herb. Her grandfather had answered, "Just watch really closely."

There were many further meetings with Johanna Paungger, and it was a long while before we finally sensed with certainty that the time was now ripe to start on a book. We had got to know one another and gained each other's trust. More and more people were coming to her lectures, beginning to show an interest in this ancient knowledge and urging her to write everything down. This book is the outcome of harmonious teamwork, a mutual co-operation that I can only describe as happy. Frau Paungger contributed her knowledge and her experience and I provided my pen and my experience. From time to time, however, you will find short passages written in the first person which depict quite personal experiences and observations by Frau Paungger or myself.

Even writing itself became a very special learning process for me. In the beginning, I forgot the old proverb that "enthusiasm for the learner is like sleep for the hunter." Gradually it became clear to me that Johanna Paungger doesn't want to prove anything and doesn't want to teach anybody, and that knowledge has no need of any kind of justification, since it proves itself solely by means of itself. What matters most to her is to dissuade the reader that he has found yet another patent remedy, perhaps even a panacea, with which to deal with all his problems. Even keeping to the 'right moment' does not help in the long run, if thought and attitude are not in harmony. A crutch, such as instilling in oneself and keeping to rules and laws, can only fulfil one purpose. One leans on it as long as it is necessary, and then one throws it away when it is no longer needed. This knowledge should become second nature and lead to alertness towards both oneself and the environment. Daily experience and experimentation with these rules sharpen our attention to our surroundings and lead us to recognise the connections with our own lives, which ultimately go beyond the rules.

In olden times, it was the noblest duty of a man of knowledge, whether craftsman or philosopher, to pass on his knowledge (not his hunches, assumptions, opinions or convictions) in a responsible manner. Now for the first time the knowledge of lunar rhythms, insofar as this is definable in writing, is available to us, together with an abundance of tips and advice dealing with almost all the important areas of our daily life, from medicine to housekeeping and nutrition, to gardening and work on the land.

Patience is the only price you have to pay in order to profit by this book. Then it can really become a building block for another world.

Thomas Poppe

The Seven Impulses
of the Moon

It is so pleasant to explore Nature and oneself at the same time,

doing violence neither to Her nor to one's own spirit,

but bringing both into balance in gentle, mutual interaction.

GOETHE

Past and Present

For thousands of years man lived largely in harmony with the manifold
rhythms of nature in order to ensure his survival. He observed the world
around him with eyes wide open and bowed to the laws of nature, at first without even questioning their purpose.

It was not merely the state of the universe that human beings observed
closely, but the relationship between that state and the actual moment of observation – the time of day, month and year; the position of the sun, moon and stars.
Many buildings of archaeological significance, dating from ancient times, testify
to the important status that our forefathers attached to the precise observation
of heavenly bodies and the calculation of their paths. Nor was this merely out
of a dispassionate scientific curiosity, but rather because by these means they
were able to derive the greatest possible benefits from a knowledge of the prevailing influences of the current grouping of stars. The calendars which they

calculated according to the course of the moon and the sun, predicted certain forces, known as impulses, which have an effect on nature, man and beast at certain times and which recur at regular intervals. In particular, they were able to predict those forces which, keeping pace with the course of the moon, exert an influence on all living things, and which contribute to the success or failure of hunting, the harvest, storage and healing.

The naturalist, Charles Darwin, in his classic work *The Descent of Man,* wrote about one of the discoveries which had been vouchsafed to countless generations before him and which had been of great benefit to them: "Man is subject, like other mammals, birds, and even insects, to that mysterious law, which causes certain normal processes, such as gestation, as well as the maturation and duration of various diseases, to follow lunar periods."

Sharpened senses, alertness, perceptiveness and careful observation of nature made of our forefathers 'masters of the art of timing'. Through observing the universe they made a number of discoveries. They noted how numerous natural phenomena such as the tides, meteorological events, giving birth, women's menstrual cycle and many other phenomena are related to the movements of the moon. They observed how the behaviour of many animals depends on the position of the moon; that birds, for example always gather the material for their nests at particular times in the lunar cycle, so that the nests dry out rapidly after a fall of rain. They saw how the effect and success of countless everyday activities, such as chopping wood, cooking, eating, cutting hair, gardening, putting down fertiliser, washing clothes, using medicines and even surgical operations, are subject to rhythms in nature. They witnessed that sometimes operations and doses of medicine administered on certain days can be helpful, while on other days they can be useless or even harmful, often quite regardless of the amount and quality of the medication or the skill of the doctor. They also observed that plants and their component parts are exposed to different energies from day to day – knowledge which is crucial for successful cultivation, tending and harvesting of crops. And that herbs gathered at certain times contain incomparably more active agents than at other moments.

In a word, the success of an intention depends not only on the availability of the necessary skills and resources, but also decisively on the timing of the action.

Understandably, our ancestors were at pains to pass on their knowledge and experience to their children. To do this, it was necessary to give the influences they observed handy, easily understandable names, and above all, to couch them in a plausible system that would enable the description of the forces and in particular the prediction of future influences, to be understood, in every time and place. A very special clock had to be invented.

The sun, moon and stars provided an external framework – the hands and dial, so to speak, of this clock, for a very simple reason that the essence of rhythm is repetition. For example, if one observes that the most favourable time for sowing a particular plant lasts for exactly two to three days each month, and that the moon is always passing through the same stars at that time, then the idea occurs to group these stars into a 'picture' and give the constellation a name that typifies the quality of the influence in question. This constellation then becomes a figure on the clock-face of the firmament.

In broad terms, our forefathers isolated twelve impulses, each possessing a different quality and colouring. To the stars through which the sun passes in the course of a year, and the moon in the course of a month, they gave twelve different names. Thus the twelve constellations of the zodiac came into being: Aries, Taurus, Gemini, Cancer, Leo, Virgo, Libra, Scorpio, Sagittarius, Capricorn, Aquarius and Pisces. Man had created an astrological clock, from which he could read the influences that were prevailing at the moment and by means of which he could work out what helpful and hindering influences the future held in store for his intentions.

Many calendars in the past were based on the course of the moon, because the forces indicated by the position of the moon in the zodiac are of far greater importance in our daily lives than those indicated by the position of the sun. Even today, many of our holidays are based on the position of the moon. For instance, Easter, since about the end of the 2nd century AD, has always been celebrated on the first Sunday after the first full moon of spring.

Towards the end of the nineteenth century, the knowledge of these special natural rhythms lapsed into oblivion overnight. Perhaps because any kind of systematisation contains within it a sort of soporific drug. If my watch tells me it's twelve o'clock, there's no longer any need for me to observe the sun. When we are no longer aware of how the impulses and forces governing a day are influenced by natural rhythms, then the rules and laws rooted in observing the world in this way rapidly lose their force.

However, the main reason why this knowledge was abandoned is because modern technology and medicine promised us faster solutions for all the problems of daily life. In a very short time, they managed to create in us the illusion that they were able to fulfil this promise. Almost at a stroke, the observation of and respect for natural rhythms seemed to have become superfluous. Finally, the knowledge lived on only in a few isolated areas.

The young farmers, foresters and gardeners of modern times laughed at their parents and grandparents, spoke of superstition and began to rely almost entirely on the extravagant use of machines and instruments, fertilisers and pesticides. They believed they could ignore their parents' knowledge of correct timing, and for a long while mounting yields seemed to justify this belief. Thus they lost contact with nature and began, at first unconsciously, to contribute to the destruction of the environment, always supported by an industry that understood how to maintain their confidence in its ability to solve every problem. Today there is hardly anyone left who can close his eyes to the high price that we have had to pay for disregarding the rhythms and laws of nature. Yields are sinking and pests are having a field day, because the soil is being exploited without being allowed to protect and regenerate itself. The use of pesticides has increased many times within a few decades without any appreciable success. The quality and health value of the harvested produce convey a very plain message.

The progress of the chemical and pharmaceutical industry has seduced the medical profession into the firm belief that they can disregard with impunity the cyclic flow and wholeness of life. The rapid removal of pain and other symptoms was enough to count as successful therapy. Research into the causes of illness and preventive measures, and the value of having the patience and willingness to create a long-term relationship with the patient receded into the background.

All of us who so lightly turn our backs on the knowledge of lunar rhythms do so because we have elevated short-term comfort to the highest good, at the expense of reason and moderation. We believe we can outstrip everything, including nature, and in the process we outstrip ourselves. In the infernal tempo of our age we rush frantically from past to future. The present moment, the only point at which life actually occurs, is utterly lost.

We have lost touch with nature and with our intuitive abilities. This loss is summed up in the following quotation:

> "The citizen is increasingly dependent on services over which he is unable to exert any influence and on experts who advise and prescribe how he is to live. His normal, inborn faculties are smothered in this welter of instructions and advice; he remains as subordinate and bereft of autonomy as a child, and is permitted – is expected – to remain so. He has no confidence in himself, in the future or in the self-regulating power of life."
>
> Ricarda Winterswyl, *Süddeutsche Zeitung*, Munich, 20.4.1991

On the other hand, we may ignore these rhythms for the simplest of possible reasons. It is possible that we may have no knowledge of them. Perhaps you are one of the pioneers who are trying to reconquer this knowledge, slowly, unhurriedly, little by little. For it is by no means too late to revive this ancient art. It is merely waiting for people who don't make excuses that 'on your own you can't achieve anything'. Even if at the present time there are so many indications that the individual can have no influence on the recovery of our environment, every single action counts, be it ever so small. Sometimes it can count for much more than grand gestures and above all grand words.

All the rules and laws presented in this book have their roots in personal experience. There is nothing that derives from hearsay or is based on assumptions.

In the following pages, the seven different states of the moon are described.

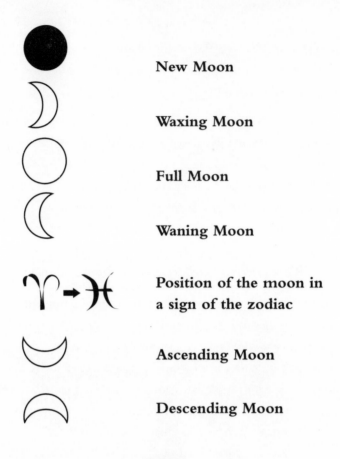

New Moon

Waxing Moon

Full Moon

Waning Moon

Position of the moon in
a sign of the zodiac

Ascending Moon

Descending Moon

The New Moon

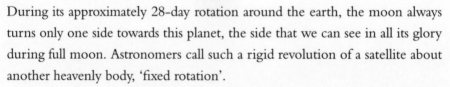

During its approximately 28–day rotation around the earth, the moon always turns only one side towards this planet, the side that we can see in all its glory during full moon. Astronomers call such a rigid revolution of a satellite about another heavenly body, 'fixed rotation'.

If the moon lies, as seen from the earth, between the earth and the sun, then the side turned towards us is completely in darkness. Then it cannot be made out, and on the earth **New Moon** prevails, also known in olden times as the 'dead moon'. See diagram on page ii.

It is important to note that at new moon, the moon stands in front of the same backdrop of stars and hence in the same zodiac sign as the sun. This is understood easily when one considers that the moon at this time is at its nearest to the sun and thus sun, moon and an observer on earth almost form a single line. So for instance, in March the new moon is always in the sign of Pisces, in August it is always in Leo, and so on.

It is useful to bear this rule in mind, when trying to work out roughly which zodiac sign the moon is currently in. Remember that the moon always stays for two or three days in a particular sign. So by the next full moon after the new moon in March, the moon has covered exactly half the zodiac and is thus six signs further on. According to this rule, fourteen days later the moon will be in the sign of Virgo or Libra. This principle can be applied to every other month in the year.

In calendars, the new moon is usually depicted as a black disc. A short period prevails in which special impulses affect humans, animals and plants. Anyone who fasts for a day during this period can prevent many illnesses, because the detoxifying power of the body is at its peak. If someone wishes to throw bad habits overboard, then this day is more suitable as a starting point than almost any other. The optimum moment to renounce bad habits is the new moon in March, when the sun changes from Pisces to Aries. However, as this is only an annual event, it is not worth waiting for months to make your good resolution. Diseased trees recover after being cut back on this day. The earth begins to breathe in.

The impulses prevailing on new moon days are not so strongly and directly perceptible as those of the full moon, because the switch-over and reorientation of the forces from waning to waxing moon is not as violent as that from waxing to waning moon.

The Waxing Moon

Just a few hours after new moon, the side of the moon facing the earth begins to come into view. Moving slowly from right to left, a graceful sickle appears. In its turn the **waxing moon,** with its own specific influences is getting under way. The journey of about six days to the half-moon, is known as the first quarter of the moon. The period from there until full moon, about thirteen days later, is called the second quarter.

The influences of the waxing moon are both favourable and unfavourable. Everything that is to be supplied to the body – that builds it up and strengthens it – works twice as well for a period of two weeks. More children come into the world when the moon is waxing and when it is full. However, the more the moon waxes, the more unfavourable are the prospects for healing wounds and for operations. On a more domestic note, even with the same amount of detergent, washing does not become as clean as it does when the moon is on the wane.

The Full Moon

Finally the moon has covered half of its journey round the earth. The side facing us appears as the **full moon**, a bright circular disc in the night sky. Seen from the viewpoint of the sun, the moon is now behind the earth. On calendars, the full moon is depicted as a white disc.

In the few hours of full moon, a clearly perceptible force can be felt on earth among humans, animals and plants. 'Moon-struck' people walk in their sleep; wounds bleed more profusely than at other times; medicinal herbs gathered on this day develop greater powers and trees cut back at this time could die. Police stations increase their manpower, because they regularly face a rise in violent crime and accidents, and midwives lay on an emergency shift.

The Waning Moon

Slowly the moon wanders onwards; the shadow dents it from right to left, as it were. The roughly thirteen day phase of the waning moon begins, known as the third and fourth quarter.

Once again, our forefathers deserve the credit for the discovery of special influences during this period. Operations are more successful than at other times; almost all work about the house goes more smoothly; even someone who eats too much at this time will not put on weight so quickly. Many jobs in the garden and out of doors are either favoured at this time, such as sowing and planting root vegetables, or else they turn out rather unfavourably, such as grafting fruit trees.

The Moon in the Zodiac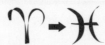

When the earth travels round the sun, the latter, seen from the viewpoint of the earth, stays for a month in each of the constellations of the zodiac. The moon passes through the same zodiac signs in its 28-day orbit of the earth, though in its case, it only stays about two and a half days in each sign.

The twelve different forces that are associated with the position of the moon in the zodiac, can only seldom be perceived so directly as the force felt at full moon. However, the influence of these forces on plants, animals and humans is clearly recognisable, particularly in their effects on the body and health.

Lunar rhythms also affect life in the garden and on the farm, influencing crop yields, weed control, and fertilisation. For example, the moon in Virgo, whose element is earth, is regarded in the plant kingdom as a 'root day'. Measures taken to enhance root growth during these two or three days are more effective than on other days.

In the field of medicine, physicians used to acknowledge the connection between the position of the moon and the course of illnesses. Hippocrates, the mentor of all physicians, knew about the lunar forces and instructed his pupils unequivocally: "Anyone who practises medicine without taking into consideration the movement of the stars is a fool." He also advised: "Do not operate on parts of the body that are governed by the sign through which the moon is passing."

In human beings, the current position of the moon in the zodiac exercises specific influences on regions of the body and its organs. One commonly speaks of each bodily zone being governed by a particular sign of the zodiac. The table summary at the end of this chapter illustrates the connections between the signs of the zodiac and bodily parts.

Those of our ancestors skilled in the art of medicine made a number of discoveries regarding the link between the zodiac and parts of the body. They found that everything that is done for the well-being of that part of the body governed by the sign through which the moon is currently passing, is more effective than on other days, with the exception of surgical operations. They noted that everything that puts a special burden or strain on that part of the body governed by the sign through which the moon is currently moving, is more harmful than on other days. They also observed that operations on the organ or body part in question should be avoided during these days, if at all possible. Emergency operations are subject to a higher law.

If the moon is waxing as it passes through the zodiac sign linked to a particular part of the body, then all measures taken to supply nutrients to and strengthen the region of the body governed by the sign, are more effective than when the moon is on the wane. On the contrary, if it is waning, then all measures taken to detoxify and relieve the organ in question are more successful than when the moon is waxing.

The only obvious exception to this rule is surgical operations. Admittedly, they ultimately serve the well-being of the organ in question. However, at the moment of the operation and in the period that follows immediately after, their effect is to put a strain on the organ. These relationships will be discussed again in detail in chapter two.

The Ascending ☽ and Descending ☾ Moon

We shall be referring to the **ascending** and **descending** forces of the moon frequently. The important thing to realise is that whether the moon is ascending or descending has nothing to do with the phases of the moon – whether it is currently waning or waxing. The ascending and descending moon is a concept connected with the position of the moon in the zodiac.

All signs of the zodiac through which the sun passes in the course of a year, between the winter solstice on December 21st and the summer solstice on June 21st, (from Sagittarius to Gemini), possess an **ascending** force – the force of winter and spring, which signals gradual increase, expansion, growth and flowering. On the other hand, a **descending** force, belongs to the signs in the second half of the year (from Gemini to Sagittarius) – the forces of summer and autumn, which signify ripeness, harvest, decline and rest.

However, the signs of Gemini and Sagittarius differ from the others, because they represent turning points between the ascending and descending forces and for this reason it is not possible to ascribe them definitely to one or other of the two forces.

Sagittarius			Gemini		
Capricorn			Cancer		
Aquarius			Leo		
Pisces	}		Virgo	}	
Aries		Ascending	Libra		Descending
Taurus		force	Scorpio		force
(Gemini)			(Sagittarius)		

Both these qualities, ascending and descending, can be felt during the 28-day journey of the moon through the zodiac, almost as if the forces of spring, summer, autumn and winter were also perceptible in the course of a single month. They contribute to the individual colour of a given sign and exercise an effect – depending on the phase of the moon – particularly in the garden and out of doors, but also on our health.

The period of the ascending moon was called 'harvest time' and that of the descending moon, 'planting time'. When the moon is ascending (between Sagittarius and Gemini) the sap is rising; fruit and vegetables are especially juicy, and plants that grow above ground do particularly well. When the moon is descending (between Gemini and Sagittarius) the sap is drawn downwards and this enhances root formation.

In former times, people devised a mnemonic in order to be able to distinguish between the two impulses. When the moon-like sign shaped as a bowl ☽ is shown in a calendar this signifies an ascending moon. The bowl is being filled up, as in harvest time. Conversely, the inverted sign ☾ indicates the descending moon, ideal for planting time.

The Aries and Taurus influences are ascending. These first two signs of the zodiac govern the upper extremities of the body, from the head to the neck and shoulders. The next sign, Sagittarius, is on the cusp of the ascending and descending forces, belonging to neither, followed by Capricorn, Aquarius and

Pisces, that are all ascending and govern the lower extremities of the body, the thighs, knees, lower legs and feet. All these signs of the zodiac are directed outwards, upwards and sideways from the shoulders and downwards from the thighs.

The middle six zodiac signs, from Gemini to Sagittarius are descending and directed inwardly into the body and principally concern the internal organs – the heart, the lungs, liver and kidneys, down to the hips.

Combined Effects

The seven impulses described above are able to work together to produce a particular effect and to enhance one another, or alternatively to cancel one another out.

For example, measures taken to detoxify the body when the moon is on the wane, are more effective in a zodiac sign that has a descending force, than in a sign with an ascending force. Because all the internal organs, including those responsible for detoxification, are governed by signs with a descending force.

If, for instance, the moon is in Leo, which would have an unfavourable effect on a heart operation, this negative effect will be amplified many times if the moon is waxing at the same time, (as is the case from February to August). On the other hand, a remedy to strengthen the heart, administered in a sign with ascending force, when the moon is waxing, is more effective than the same remedy applied when the moon is on the wane, in a sign with descending force.

We leave it to the pioneers among our readers, in particular, the physicians and complementary medicine practitioners, to study the interaction of the zodiac signs with the waning and waxing moon and to draw the appropriate conclusions.

I know a few doctors and practitioners personally, who have been building up experience with these rhythms. Even rather protracted illnesses can often be relieved using this method and with a slight shift of attitude. Observation and patience is called for here. Neither is especially suited to our fast-moving age.

The Occurrence of the Zodiac Signs in the various Phases of the Moon

Zodiac Sign	in Waxing Moon	in Waning Moon
Aries	October to April	April to October
Taurus	November to May	May to November
Gemini	December to June	June to December
Cancer	January to July	July to January
Leo	February to August	August to February
Virgo	March to September	September to March
Libra	April to October	October to April
Scorpio	May to November	November to May
Sagittarius	June to December	December to June
Capricorn	July to January	January to July
Aquarius	August to February	February to August
Pisces	September to March	March to September

For quick reference, you can make a note of the following rule of thumb. The sign of the zodiac in which the sun is presently situated, will be in the waning moon for the next six months and for the following six months will move into the waxing moon. Thus, if we take March, the sun is in Pisces, and the new moon is in Pisces.

If one combines a knowledge of the bodily regions that are governed by various zodiac signs with the information given above, concerning the occurrence throughout the year of the signs of the zodiac in the different phases of the moon, and the laws attendant on this, then it follows logically that measures taken to heal certain organs and regions of the body will have different effects in each half of the year – a conclusion substantiated by experience on numerous occasions.

For example, hip operations, (the region of the hips is governed by the sign Libra), will produce much better results in the months from October to April, when Libra is in the waning moon, than in the other half of the year

– provided that the operation does not take place precisely during the Libra days.

When one has grasped this principle, it is very easy to transfer it to other regions and organs of the body.

So every sign of the zodiac, each offset from its predecessor by a month, has at its disposal the sustaining power to flush out poisons for half the year, and to supply constructive materials for the other half. I discovered this connection because I always note down the time when a measure is particularly helpful or simply has a good effect. I also make a note whenever an otherwise 'good' remedy, administered in favourable phases of the moon or in the appropriate sign of the zodiac, fails to produce the usual effect. From observations over many years, I have come to the conclusion that certain applications in Autumn, lead more rapidly to the desired result than in spring, and vice versa.

Special Rhythms

In this book you will be introduced also to some very special rhythms – rules and dates that are completely independent of the position of the moon. They are amongst the most remarkable and inexplicable phenomena and we shall not even attempt to give reasons for them. How is one to explain the fact that wood cut down after sunset on March 1st will not burn?

We are confident that there are interested and curious readers who simply will try these extraordinary laws out for themselves. They are just as valid as any other rule.

The Moment of Contact

There is one question that interests many people. How can it be that a particularly favourable moment to gather fruit, say, or to take a medicine, often produces very positive long-term results, whereas only a short time afterwards, the opposite negative influence prevails which condemns the same action to failure? Then is it not possible for the subsequent negative energy to cancel out the positive?

Potatoes, for example, last for months if they are picked and stored at the correct moment. If they are harvested only a few days earlier or later they can spoil very rapidly sometimes.

Perhaps the answer will sound a little mysterious, although there is evidence to back it up. The moment of 'touching' is the decisive factor, the moment of action, whether it be on a human being, an animal or a plant.

If I touch a living being at a particular moment, whether with my thoughts or my hands, through my inner and outer intentions, then at that moment I transmit fine energies, and in particular the forces that are characterised by the phase and zodiac sign of the moon, like a magnifying glass which gathers the ambient scattered energies together and produces a greater effect with them combined, than if they were separate.

To stay with our example, these forces have such a powerful effect that even 'touching' potatoes that have already been stored will have a different outcome, depending on the moment of contact. If one realises that potatoes are spoiling prematurely, because they were picked and stored at the 'wrong' moment, then it is still possible to save the situation to a considerable extent, by storing them again on a favourable day. Conversely, sometimes one can discover that stored potatoes suddenly spoil very quickly if they are touched and moved at an unfavourable moment – perhaps through sliding down the pile after a few potatoes have been fetched for kitchen supplies.

Many apparently contradictory experiences in everyday life – in medicine, the garden, the countryside and the household – may be explained by the principle of contact. This may be applied to all the rules that are presented in this book.

The Lunar Calendar

The only technical aid required to master the knowledge of lunar rhythms is a lunar calendar, which is one that gives the phases of the moon and the position of the moon in the zodiac. You will find one at the back of the book.

We have received countless letters from all over the world telling us what is of particular interest to our readers in connection with these calendars and what experiences they have had with them over the years. We are now able to pass this experience on to you.

Our lunar calendar is calculated according to the position of the moon in the zodiac. All the useful experience which we pass on in the domain of medicine, medicinal herbs and ecologically sound building, in gardening, agriculture and forestry, is based on this calendar. Frequently we have received queries because some regional lunar calendars exhibit slight differences from ours. You need not worry about this, as most of these differences arise because many lunar calendars have been calculated by astrologers or astronomers according to their own principles, without any regard for lunar rhythms, which are for the most part unknown to them. If you have any doubt about the validity of a particular calendar, simply experiment with both calendars until you can be certain.

The times of the new and full moon have been left out, since these alter from one time-zone to another. For some activities, however, it is important to know the exact times of the phases of the moon and these can be found in most conventional calendars.

Anyone who studies astronomy and is able to identify the individual constellations of the zodiac in the night sky, will discover that the actual position of the sun and moon differs from that given in the calendar. However, you can place your trust in the calendar. Certain deviations in the course of sun, moon and stars over a 26,000 year cycle are responsible for this.

For this reason, for thousands of years valid lunar calendars have not been calculated according to the actual position of the sun, but from the vernal equinox, when day and night are the same length, on about March 21st. Perhaps the discrepancy between the actual position of the moon, and the position given in the calendar, can throw light on the causal relationship of lunar rhythms. It should be clear from this that the stellar constellation itself, thousands of millions of light years away, has nothing to do with our recognising and exploiting the twelve impulses. It is not that the stars exert a force, but rather that a force exists which may be calculated with the aid of the stars.

Perhaps we might offer a suggestion. A researcher wishing to fathom the reciprocal relationship between the position of the moon and the quality of the impulse, should keep a look out for resonance phenomena (the sympathetic vibration of a body with the vibration emanating from another body) arising from the revolutions and oscillations around the sun of the moon, earth and planets – as if he were studying a multi-dimensional musical instrument that produced twelve clearly distinguishable tones.

Just how important resonance is, the sympathetic vibration of a body with the vibration emanating from another body, is something I discovered as a very small child. While I was playing in the garden near a zinc bathtub, I suddenly heard some soft music for about a minute, and then the voice of a news reader. The sounds were coming from the bathtub and they stopped the moment that I grasped hold of it. Under certain conditions the tub had the same frequency as a radio transmitter in the vicinity, and the way it was constructed amplified the waves and made them audible.

It is actually possible to calculate to the exact minute, the shift from one zodiac sign to the next, but calendar-makers today, as in all previous ages only give the position of the moon in the zodiac for whole days, and for good reasons pay no attention to over-exact methods of measurement. The influences indicated by the position of the moon in the zodiac overlap and merge, particularly when the calendar shows a sign three days in a row. Then the force of the neighbouring sign can still be felt on the first day of the following sign, or can already be felt on the third day. Nature is flexible, so that for every suitable time that is missed or spoilt by the weather, sufficient alternatives are available in order to achieve results that are nearly as good.

Those people who have grown up with the lunar calendar and watch out for the right moment for their activities, often no longer need to consult the calendar, because there are numerous signals in nature that indicate the change from one sign to another, once one has begun to pay attention to them. For instance, the penetrating light on air-days that occurs in the signs of Gemini, Libra and Aquarius; improved body circulation during Leo; differences in the way windows steam up on water-days and air-days; the slight headache when Aries arrives; the digestibility of a fatty meal during Gemini, Libra and Aquarius, and many more. People with intuitive gifts, or green fingers are often guided unconsciously by all these signals, which show us the most sensible way to proceed.

Nature does not allow itself to be forced into a rigid system and governed according to a set of handy formulae, even if that is what our laziness constantly cries out for. We consider that to be one of its most beautiful and life-giving qualities. The lunar calendar is a valuable aid, no less, but no more either. It is not intended as a substitute for your personal awareness and experience. On the contrary, it can serve as a key to the enlargement of your awareness. Out of this experience an awareness can grow that will be of use in all areas of your life.

All the rules concerning lunar and natural rhythms can be applied to both the northern and the southern hemispheres. Slight exceptions between the two are due to the reversal of the seasons in the southern hemisphere. Our winter is high summer in the southern hemisphere and when in the temperate latitudes of the southern hemisphere the leaves are falling, we are being wafted by spring breezes. This difference is of special importance in gardening, agriculture and forestry. For instance, the right time for felling timber varies from the southern to the northern hemispheres. This should be done mainly when the sap is at rest, in temperate or cold regions, between 21 June and 6 July, or in tropical regions, during the period of greatest heat and aridity. With a little experimentation all the guidelines in the book can effortlessly be transposed to the southern hemisphere with its reversed sequence of seasons.

However, perhaps the most relevant difference between the northern and southern hemispheres is the external form of the waxing and waning moon in the sky. In the northern hemisphere, the moon waxes from right to left and in the southern hemisphere, from left to right. Since probably ninety percent of our readers live in the northern hemisphere, we have depicted the symbols for the waxing and waning moon as they are to be observed in the northern hemisphere – the exact reverse of what can be seen from the southern hemisphere. To make matters simpler, look at it this way. When we talk about the waning moon, we mean the period between full moon and new moon, regardless of the form manifested in the sky or depicted in the calendar. When we talk about the waxing moon, we mean the period between new moon and full moon, regardless of its form in the sky or in the calendar.

Ultimately, it comes down to your own common-sense. The information in this book can be taken first and foremost as a stimulus for the journey into the realm of natural and lunar rhythms. Build up your own experience, experiment and try things out. You will soon discover for yourself exactly what difference the 'topsy-turvy' seasons make down under.

The Zodiacal Table

The table that follows is an important tool. It gives an overview of the various effects of each individual sign of the zodiac on regions of the body, parts of plants and different types of food. Also it shows the most commonly used signs of the zodiac, in order to facilitate looking up the symbols in the calendar at the back of the book. You may wish to make a copy of it to consult while reading through the book.

Zodiac Sign	Symbol new/old	Body Zone	Organ System	Plant Part	Element	Ascending/ Descending	Food Quality	Day Quality
Aries		head, brain, eyes, nose	sense-organs	fruit	fire	◡	protein	warm
Taurus		larynx, speech organs, teeth, jaws, throat, tonsils, ears	blood-circulation	root	earth	◡	salt	cool
Gemini		shoulders, arms, hands, lungs	glandular system	flower	air	◠	fat	air/light
Cancer		chest, lungs, stomach, liver, gall-bladder	nervous system	leaf	water	◠	carbohyd-rate	water
Leo		heart, back, diaphragm, circulation, artery	sense-organs	fruit	fire	◠	protein	warm
Virgo		digestive organs, nerves, spleen, pancreas	blood-circulation	root	earth	◠	salt	cool
Libra		hips, kidneys, bladder	glandular system	flower	air	◠	fat	air/light
Scorpio		sex organs, ureter	nervous system	leaf	water	◠	carbohyd-rate	water
Sagittarius		thigh, veins	sense-organs	fruit	fire	◡	protein	warm
Capricorn		knee, bones, joints, skin	blood-circulation	root	earth	◡	salt	cool
Aquarius		lower leg, veins	glandular system	flower	air	◡	fat	air/light
Pisces		feet, toes	nervous system	leaf	water	◡	carbohyd-rate	water

Healthy Living in Harmony
with Lunar Cycles

Do you think you can take over the universe
and improve it?
I do not believe it can be done.

The universe is sacred.
You cannot improve it.
If you try to change it, you will ruin it.
If you try to hold it, you will lose it.

So sometimes things are ahead and sometimes they are behind;
Sometimes breathing is hard, sometimes it comes easily;
Sometimes there is strength and sometimes weakness,
Sometimes one is up and sometimes down.

Therefore the sage avoids extremes, excesses and complacency.

LAO TSU

Reflections on Health

Everyone thinks he knows what health means. And yet only a few years ago it was officially defined at the highest level by the World Health Organisation as 'the absence of illness' – perhaps precisely because it is only when health is lacking that one is conscious of it being the 'highest good'. In spite of this, nothing could be further from the truth. Many doctors and alternative practitioners today are making an effort to see things from a new perspective. For example, the German physician and author, Dr Harald Kinadeter, writes: "For us health means the power and capacity to become what we are and to overcome whatever stands in our way. It is a harmonious interplay of the various processes that define a human being in the service of an intention that pursues the meaning of life."

Our ancestors knew about these ideas. Priest-physicians, shamans and medicine men, skilled in the art of healing, acted on the knowledge that we humans are not merely machines. They understood that we are more than a system of bones, nerves, muscles and organs and that body, mind and soul mutually influence each other and form an inseparable union with everything around us – with other human beings, with nature, even with the stars. They knew that illness arises whenever a human being, for whatever reason, can no longer maintain the dynamic, flowing equilibrium between the many elements of life – between tension and relaxation, healthy egoism and devotion and the ups and downs of fate.

Everything in nature is a combination of sound, vibration and rhythm. A balanced life, thus means that one does not continually disregard the cyclic rhythms to which our body is subject or constantly try to swim against the tide. On the other hand, balance has nothing to do with the rhythm of the clock, with laziness and a sluggish, lukewarm ebbing away of time. Measured doses of excess are just as important for healthy living as regularity and rhythm in daily life. Every organ, every living creature from time to time requires its share of stimuli, in order to push forward to the limits of its potential.

Our body is like a robust and dependable ship, which requires a certain degree of maintenance in order to deliver its full performance, or the performance corresponding to its age and which needs a more or less regular supply of fuel, in the form of oxygen and a varied diet. It is not only the body that needs food, but also the mind and the soul. The scientific view of the world that dominates the modern mind has made us forget that the vessel that comprises the body, our feelings, thoughts and instincts also has an important part to play and that, most important of all, it is waiting for a captain – consciousness.

It is our consciousness, our attitude towards life, which helps to determine the fate of our body and exerts an influence on health, efficiency and the enjoyment of life. At the same time, it affects the fate of the natural environment. The state of the latter is always an exact reflection of our own state, physically as well as mentally.

Our body is a truly marvellous thing. For years on end it appears to forgive everything – faulty nourishment, lack of exercise, excess intake of alcohol and nicotine, stress and long-standing disregard for its natural rhythms. So much has become cherished routine that a change in our daily habits is difficult to achieve. We have grown out of touch with the true requirements and rhythms of our body.

In order to become an experienced and benevolent manager of your own life, you have to know how your body functions, how much you can abuse it, and under what conditions it will perform to its full potential. Great courage is needed for this. The courage to recognise that one almost always reaps what one has sown, that illness simply does not strike us down out of a clear blue sky, and that our whole life is subject to cycles, with peaks and troughs, highs and lows.

In the West, we are living through a reign of 'fitness terror'. We are under constant pressure to be continually fit, beautiful, on top form and alert. This ideal is almost a declaration of war on nature, for nature also imposes on us troughs in the waves, and we would do well to learn to like them. It takes courage nowadays to be able to face set-backs with a cool head and calm resignation. Common sense is a great leveller. Anyone who has learnt to listen to it knows that "There are no hills without valleys."

> *Man cannot have good times for a thousand days,*
> *any more than a flower can bloom for a hundred days.*
>
> TSENG-KUANG

Where previously a calm weighing up of possibilities and alternatives determined one's way of life, nowadays the principle of comfort holds sway – comfort and apparently limitless possibilities. Whether it is our health that is in danger, or the health of nature, the law of reason, not comfort, ought to come first. We do not need, nor ought we, to renounce progress and technology. But our watchword should be moderation in all things. Then a rational life and an intact environment can be achieved while making full use of research and technology.

If we manage to show even a single reader just how easy and enjoyable it is to submit to the rhythms of nature, out of friendship towards oneself, then we shall have achieved a great deal. Only someone who is his own best friend can be a friend to his fellow men.

A Few Important Rules of Health

The greatest wisdom reveals itself in the simple
and natural arrangement of things,
and people do not recognise it
precisely because everything is so simple and natural.

JOHANN PETER HEBEL

In this chapter you will encounter a few more rules for health connected with the lunar cycles. They are based, like all the other rules, on first-hand experience and long years of observation. If you are looking for proofs, you will have to try things out for yourself, calmly and patiently. This is the only proof that can be offered and that has any validity. Doctors and statisticians, however, will not have much difficulty checking the correctness of these rules if they use their patients' records to compare the varying success-rate of cures with the cycles of the moon. The reaction, "I've never heard of such a thing, so it cannot be true," may not be very helpful. However, it is understandable enough. Only a few hundred years ago people were tarred and feathered for maintaining that silk comes from caterpillars. Everyone 'knew' that silk was made by angels.

In modern times, it is without doubt, difficult to adapt to living in harmony with the rhythms of nature. Almost all events, rituals and customs in private and professional life no longer make allowance for the inherent impulses of nature. This is how stress arises in its manifold forms, so often compelling us to forget and ignore nature's signals, natural instinct and common sense.

A great deal could be gained by recognising that health-damaging stress, is in the majority of cases, created by ourselves. It is frequently the consequence of too much or too little volition at the wrong moment. It often arises, when inwardly or outwardly one is not up to a self-imposed task, or if one is inwardly resisting it.

Our body reacts if we continually force it to ignore its natural rhythms and needs. Not at first, when we are young and can shake off negative effects like water off a duck's back. But gradually these many little impulses add up until they lead to an illness – the tip of the iceberg – whose cause can only be traced with great difficulty. To turn a German saying on its head: "What lasts a long time finishes ill."

This book is not a cure-all, full of instant remedies. The effects of going against natural rhythms are slow to make themselves felt and likewise living in tune with these rhythms will be slow to show positive results. By sitting back quietly, once a day and briefly reflecting which activities in daily life can be harmonised with the lunar rhythms, then solutions are bound to arise. Simply observing the world about us will allow us to live in harmony with nature. Progress can be made gradually, by focusing on the long-term.

There is at least one thing you can do for a start. All strenuous daily work and hobbies (even the latter have for some people degenerated into hard work) which do not have to be done at a specific time, should be postponed for a while until the moon is in its waning phase. Act with caution, closely observing the consequences.

Nothing is more convincing than your own personal perception. Gradually you will become aware of how pleasant it is to work in harmony with the lunar rhythms; to no longer restrain your strength when the moon is waning and to slow down, gather your energies, prepare and plan when it is waxing. You will then begin to wonder how you were able to manage for so long without using this knowledge and why you had never noticed it before.

CAUSE AND THERAPY

Living in harmony with lunar rhythms can help us to overcome ill-health. However, to achieve true well-being we have to confront the causes of illness. Without understanding the roots of ill-health, any treatment we turn to will be limited. By taking responsibility for ourselves and acknowledging that we can play an active role in achieving good health, we can assist the healing process. When you have found an honest answer to the cause of your illness, a doctor will be in a much better position to help you. Doctors can only help you to help yourself and to awaken your own powers of self-healing. If deep down in yourself you haven't the slightest interest in getting well, because illness brings you attention and allows you to evade other responsibilities, then no doctor will be able to help you.

The causes of illness are often not to be sought in our bodies, but in the breeding ground of the destructive habit of thought, often connected with competitiveness, anxiety and greed. This is often the real cause underlying the 'many sins against nature' that the ancient Greek physician, Hippocrates mentions.

Anyone who casts a cool, objective glance in the mirror will not be able to reject these weaknesses straight away. It is this honest look at oneself, that will enable you to make the best possible use of the tips on the following pages.

It must be pointed out that this book can never replace and is never intended to replace the doctor. No one should treat an illness single-handedly without the advice of a physician.

At my lectures people sometimes tell me about good and not-so-good doctors. I think that there are only good doctors, but there certainly are some that are less successful with some treatments. The correct timing of the start of a course of treatment is often decisive. At first, of course, in conversations with doctors about the moon and its influence I sensed a deep mistrust. But every one of them that took the trouble to check back in his records and compare successful and unsuccessful recoveries with the position of the moon at the time of recovery, inevitably was amazed. Now, thank goodness, there are many doctors who, when treating patients, pay attention to the lunar cycle. In the case of recurrent or chronic clinical pictures, there is a particularly good chance of picking out favourable moments for treatment.

Before we come to the specific influences of the lunar phases and the position of the moon in the zodiac, it might be useful to describe once again two important phases of the moon in connection with a healthy life-style and diet:

The waxing moon

supplies, plans, takes in, builds up, absorbs, breathes in, stores energy, gathers strength and is conducive to rest and recovery.

The waning moon

washes out, sweats and breathes out, dries and is conducive to action and the expenditure of energy.

If you are aware of the differing effects of these two moon phases then you will already have taken a great step towards integrating the lunar cycles harmoniously into your daily life. You don't have to take this simply on trust. Observe and investigate – you will be able to recognise these influences yourself.

THE INFLUENCE OF THE POSITION OF THE MOON ON DIET

Good digestion is of central importance for a healthy life. Many minor and major disorders arise from imbalances in diet and digestion. Whether or not a meal agrees with us frequently depends on the position of the moon. When the moon is waxing, even if we eat the same type of meals and similar quantities, we feel full much more frequently and gain weight more easily, than when the moon is on the wane. Conversely, when the moon is waning, one can eat more than usual without immediately putting on weight.

Diet and digestion are not only influenced by the phases of the moon, but also by the position of the moon in the zodiac. This knowledge is still ignored by nutritionists and has all but disappeared from our awareness.

The following table describes the relationship between the position of the moon in the zodiac and the food quality of a given day.

Warmth days   *Element:* fire   *Plant part:* fruit

Aries
Leo   } *Food quality:* protein
Sagittarius

These days have the best **protein qualities**. Protein has special effects on the physical body and the sense organs.

Cool days   *Element:* earth   *Plant part:* root

Taurus
Virgo   } *Food quality:* salt
Capricorn

It is here that the best **salt qualities** prevail, which are favourable for nourishing the blood.

Light days   *Element:* air   *Plant part:* flower

Gemini
Libra   } *Food quality:* fat
Aquarius

These have the best **fat and oil qualities** and they supply the glandular system.

Water days   *Element:* water   *Plant part:* leaf

Cancer
Scorpio   } *Food quality:* carbohydrate
Pisces

These days possess good **carbohydrate qualities** and influence the nervous system.

The changing impulses of the moon as it passes through the zodiac, affect our food and the ability of the body to utilise it. For instance, when the moon is in the air signs of Gemini, Libra and Aquarius, good fat and oil qualities prevail. During this period, the oil contained in an olive behaves in a special way, differently than on other days. For example, it is possible to extract much more oil from the fruit than on any other day and the capacity of our body to make optimal use of this oil also changes. In other words, the harmonious interplay between food, plant and body is also dependent on the timing of the meal. During the water days of Cancer, Scorpio and Pisces, carbohydrate qualities are predominant and bakers may observe that their shelves often empty much more quickly during these days than usual.

There are no hard and fast rules regarding these food qualities. For instance, some people can digest bread particularly well on the water days of Cancer, Scorpio and Pisces, while others get a bloated belly after two slices. It is best to experiment with the charts and observe for yourself how the different food qualities influence your health. With patience, after only a few months, or even weeks, using this table together with the lunar calendar, you will be able to determine exactly which food is particularly good or bad for you and on which days.

- If, for example, your glandular system is slightly disturbed, then you should pay close attention to what foods are particularly tasty on air days in Gemini, Aquarius, and Libra. You may discover that precisely those foods least suited to you taste especially good, and that a slight adjustment is required in your menu. It is easier to go without certain foods for a few days in the month, than to keep to a strict dietary regime. This is how you can proceed on other days, too.
- If you are especially fond of eating bread or other flour-based foods on water days in Cancer, Scorpio and Pisces and have problems with your weight, then you should try eating easily digestible types of bread on these days and should steer clear of foods with a high carbohydrate content, such as pasta.
- Earth days in Taurus, Virgo and Capricorn, have an especially strong influence on the salt quality. On these days it would be better to avoid large

quantities of bacon, ham, salted herrings, fatty cheese and the like. Many people absorb salt in especially large quantities precisely on these days, so they have to be doubly careful at this time. If your doctor has prescribed food that is low in salt, then these days are particularly tricky. Unfortunately, on precisely these susceptible days you may have a special craving for salt. By following the motto 'just this once' the good effects of a whole month of abstinence from salt can be completely ruined. Observation will gradually show the way and make you better prepared for these days.

On warmth days in Aries, Leo and Sagittarius, you should observe whether your menu contains a noticeably large or small quantity of protein or fruit, and what effect this has on you. Warmth days are also fruit days, because the fruit part of the plant is especially favoured at this time.

Of course it is difficult to make observations and draw conclusions when you are not cooking for yourself. But even when you have ready made meals you can establish whether something tastes good, or whether the food lies heavily on your stomach afterwards. This will provide valuable insights for when you are able to decide on your menu for yourself. Any change in a positive direction, however small, counts for something.

It is also valuable to notice varying reactions to the same food. Does it happen that something fatty is good for you one day, but not a week later, and then later it tastes really good again? Take a look at the moon calendar and make a note; and then in the course of time draw your own conclusions.

If on fruit days in Aries, Leo and Sagittarius, you only feel like eating protein foods or fruit, or if on root days in Taurus, Virgo and Capricorn, you help yourself to salty food, then this isn't a bad rhythm, as long as it does you good. There is no need to labour the point that doing good and tasting good are often two different things. For people who are allergic, careful observation of when a particular food triggers off an allergic reaction can be very beneficial. The food causing the allergy is not equally harmful on all days. With the aid of a moon calendar one can easily find out what sort of an influence a particular day has on the allergen.

In general, if you take into account the four food qualities in the course of the month, giving the current quality greater emphasis and building it 'one-sidedly' into your diet, then you can't go far wrong. At the very least, you will rapidly establish whether you are one of those for whom this rhythm applies. For many people, however, the currently prevailing quality is precisely the one they cannot tolerate, and they have to reduce their intake of foods with this quality.

Observe, be aware and take notes. Experience is what counts, not this book on its own. It is only intended to serve as an aid for you.

FASTING

Fasting (a short or longer period without solid food) has become fashionable in recent years and can be very beneficial when it is intended as a health cure. Going without food for a brief period gives the body an opportunity to detoxify and regenerate itself and often has an extremely positive effect. However, as a means of losing weight, it is almost never successful in the long-term.

Carnival is followed by Lent, a fasting period determined by the moon. Exercising moderation in one's eating during this period is very good, because the body detoxifies and regenerates itself particularly well at this time. It will reward such a measure with increased powers of resistance and greater well-being.

Less well-known is the Advent fast (from the 1st Sunday in Advent until December 24th) which is likewise a very favourable time to live a little more abstemiously. Nowadays of course this is harder to achieve, but perhaps this information nonetheless can be of use to you. One thing is certain in all events. Cakes and pastries eaten before Christmas are much more fattening than the goodies we munch during the Christmas period itself.

In general, it is also advantageous to eat less in the days before full moon and at full moon itself. In any case, many people quite unconsciously eat more when the moon is waning, than when it is waxing, and this does not make them any fatter. A day of fasting at new moon can prevent a number of illnesses. On this day, the body detoxifies itself especially effectively. The ingestion of food can slow down this process or stop it altogether.

The Topic of Operations

Regarding all surgical operations, with the exception of emergency operations, the closer to full moon, the more unfavourable the moment. The actual day of the full moon has the most negative effect of all. Given the choice, one should operate when the moon is waning.

Everything that puts a special burden or strain on those parts of the body and organs governed by the sign through which the moon is currently moving, has a more harmful effect than on other days. Operations should therefore be avoided during these days, if at all possible.

Every surgeon can discover for himself how the lunar rhythms influence the outcome of operations, or may even have experience of the consequences of these rhythms. Complications and infections are much more frequent on such days, and healing and convalescent phases last longer. Towards full moon, there are often instances of severe bleeding that is hard to staunch.

Hippocrates (460–370 BC), the ancient Greek physician, wrote about this in his journal: "Do not touch with iron those parts of the body that are governed by the sign through which the moon is passing." What he meant was that a doctor should not carry out any surgical operations on regions of the body that are governed by the current sign of the zodiac. The influence of each sign of the zodiac on parts of the body can be found in the table in Chapter I. To take some examples, on Pisces days, operations should not be carried out on the feet and on Leo days there should be no heart operations.

You may wonder what should be done if the date for an operation falls in Leo, but at the favourable moment of the waning moon. In this case, the positive influence of the waning moon is stronger than the negative influence of the Leo days. To illustrate this, here is a list of the favourable and unfavourable influences on a heart operation:

Most unfavourable:	full moon in Leo
Very bad:	waxing moon in Leo
Bad:	waxing moon in another sign
Fair to medium:	waning moon in Leo
Good:	waning moon in another sign

The Significance of Cats

Many readers will perhaps be reminded now of horror stories about witches, who carry out strange manipulations by moonlight and then come home to their black cat from their nocturnal herb-gathering outings, riding on a broomstick.

The reason why cats are always appearing in witch stories is because formerly people used to observe cats and draw useful conclusions from their behaviour. Cats prefer to sleep and rest on 'bad spots' – places in the house that have negative energy (such as water veins and places where telluric rays can be detected). In former times, wherever a cat felt at home one would never set up a workplace, even less a bed.

Thus it can be seen that many a fairy tale from olden times has its roots in concrete and sensible facts.

An Important Health Factor: The Right Place

In all houses and flats there are good and bad places, regardless of what is in these places – whether it is a wall, table, chair, bed, or kitchen work-top.

The capacity to distinguish good and bad places from one another is developed to varying degrees in each person. As a rule, however, this sensitivity is greater in infancy and youth than in adults. Each individual's reaction will be different. Some people spend years sleeping on a bed that is on a 'bad place' without getting ill, while others become restless and nervous after sitting on a bad spot for a few minutes.

It has yet to be established exactly what it is that determines the quality of a place. Certainly radiation of various types is involved – subterranean water veins, telluric rays and the like. On the other hand, there also seems to be little inclination on the part of orthodox science to make a thorough investigation of phenomena that, given a modicum of good will, can be proven with the greatest of ease. As long as we have a situation where any scientist instantly risks losing his reputation the moment he turns his attention to this subject, things will stay like this for some time to come. On the other hand, scientific research into this subject using conventional methods is perhaps not appropriate. One of the most important reasons for this is that experienced dowsers and diviners are unlikely to make themselves available for such enquiries. They know that the presence of even one person who has doubts about this art and its value is enough to distort the results of a survey. Consequently in fact the only dowsers that offer their services to science are those who, to put it mildly, overrate their own capacities somewhat.

Formerly the phenomenon of the right place was well-known. In ancient China no house was ever built before the ground had been minutely investigated. The Chinese were also the first to set down in writing their experiences regarding this knowledge. Our ancestors used a whole array of tools to identify good and bad spots – divining rods, pendulums etc., as well as closely observing the plant and animal world. From this they discovered that many animals have an innate ability for detecting the quality of a place. For instance, cats, ants and bees are 'ray-seekers'. They prefer places that are bad for human beings. Ants and bees always build their nests at the intersection of two water veins. The fact that storks and swallows are widely held to bring good luck is perhaps connected with the fact that they only nest where the surroundings are largely radiation-free. Birds and also dogs, horses and cows belong to the ray-fleeing group. Wherever they settle is also a good spot for us.

Many parents will have noticed that some babies twist and turn in bed, cry a great deal and often end up in the morning lying in a corner of the cot. Many children often cannot stand being in their own bed at night. They prefer to slip into bed with their parents or with a brother or sister. School children who sit for a whole year in the same bad place often lag behind in their work without any apparent reason; whereas up to that point everything was going well. Sometimes parents set up a very expensive and tasteful children's room with a desk in it, only to find that the child still comes to the kitchen to do his homework. Generally this is a first sign that the child's desk is in an unfavourable place.

In the adult world, too, many things can be observed that are connected with good and bad places.

Formerly it was known that in certain farms, no milkmaid could stick it longer than a couple of months, that the farmer or the farmer's wife always died young there, almost as if there was a curse weighing upon the house. In most cases, these events were due to bad sleeping and work places.

Perhaps you have noticed how some housewives stand at an angle or at a distance from their kitchen table, or that there are chairs in the living room that inexplicably remain empty, or that there are certain places in your apartment where you regularly become tired or restless. Many an otherwise able and popular teacher suddenly gets a 'bad' class, with which he simply cannot get along – 'cannot get on a firm footing' – because his chair is in the wrong place. Sometimes he unconsciously copes with the situation by walking up and down or continually sitting on the desk, instead of behind it.

Unfortunately our feelings concerning a particular place are not always a sure sign of its quality. Some people have become so accustomed to the negative energies of bad spots – perhaps because their bed has been over a bad spot for years – that they are actually magnetically attracted to bad places.

However, the insight that there are good and bad places and the manner of identifying them, are extremely important when it comes to our health. Sitting or sleeping for years on the wrong place can be responsible or partially responsible for lengthy, chronic illnesses, chronic headaches, tiredness and the like. To such an extent that one is forced to what is perhaps a somewhat drastic conclusion. It is verging on suicidal to come home after an operation or a protracted illness and then sleep in the same bed in the same position. Any person suffering from chronic health disorders ought to have his sleeping or working area investigated by people who are skilled in this art.

Of course you will now be asking what you yourself can do in order to determine the quality of a place? Unfortunately, there is no patent formula. If you have begun to suspect and have reason to suppose that you yourself or someone in your family may be sleeping or working in a bad place, try rearranging the furniture. Usually a distance of about two yards away from the old spot is sufficient.

For every ill a healing herb – a short herbal

Medicinal herbs are miniature power stations. There is hardly a bodily ailment nor illness that cannot be relieved or healed by the leaves, flowers, fruit or roots of a naturally occurring herb, provided that the patient approaches the remedy and the sickness in the correct frame of mind.

Anyone who makes wise use of herbs in the kitchen is not only doing a great deal to improve the taste of the food, but is also helping to prevent many illnesses. Preventive measures to protect our health are extremely valuable. Perhaps the time has come to return to the preventive principle of the ancient Chinese who provided for the support of their doctors, in money and natural produce, only as long as they remained healthy. If a lamb in the doctor's flock fell ill, then he was released from this communal duty. In those times doctors earned their living from the health of those in their charge not their illness.

A great variety of medicinal agents are contained in herbs, often in one and the same plant. These include:

Mucilage (pectin) used to treat inflammations and to heal wounds, contained in plants such as comfrey, coltsfoot, cowslip, cornflower, daisy, deadnettle and marigold.

Essential oils with a variety of effects (disinfecting, stimulating blood supply, stimulating or inhibiting secretions, dehydrating), contained in herbs such as thyme, garlic, yellow gentian, sage, camomile, laurel, balm, fennel and basil.

Saponins, some of which have a hormonal activity and others which are expectorants (stimulating coughing and clearing phlegm), contained in, among other plants, liverwort, cowslip, birch, speedwell, heartsease and mullein.

Silicic acid for the treatment of inflammations of the skin and the mucus membrane and used to strengthen connective tissue, contained in herbs such as horsetail, heather and the stinging nettle.

Bitter substances (alkaloids, glycosides) for regulating the functioning of the stomach and intestine and alleviating tension in the stomach, contained in plants such as wormwood, yellow gentian, mugwort, dandelion, sage and marigold.

Tannic acid for inflammations of the stomach and the mucus membrane of the intestine, contained in, among other plants, bramble, lady's mantle, and speedwell.

Salicylic acid with anti-bacterial and pain-relieving effects, contained in among other plants, willow, violet, and marigold.

Styptic (blood-staunching) agents contained in among other plants, shepherd's purse, yarrow, mistletoe and dandelion.

Laxative agents which promote the activity of the bowel, contained in among other plants, senna leaves, alder bark, common buckthorn and fumitory.

This list is far from being exhaustive, and even today new agents are still being discovered. At the same time our wonder can only grow at the unerring instinct of our forefathers, who discovered the effective herb for each different illness.

Many of our culinary herbs are plants with preventive and curative powers– from parsley and chives through to rosemary, sage and lovage, to sweet woodruff and mugwort. Quite undeservedly they have now sunk to the level of taste enhancers. Perhaps even more surprising, however, is the fact that many plants that are considered weeds, from stinging nettles to dandelions, also have a medicinal effect. For instance, there is nothing better than a blood purification treatment in early spring, using stinging nettles that have been picked at the right time. As for the power residing in the young leaves or open flowers of dandelions, all those whose complaints have been relieved or healed by this means are aware of its curative power.

RULES FOR GATHERING HERBS

When gathering herbs in the wild or harvesting them in the garden, only collect as much as you need for your current purposes, or for your anticipated

winter requirements. Respect for nature and consideration for one's neighbours demand as much. Rare herbs that are protected by law must remain
taboo.

Always restrict yourself to herbs that you know well and can identify with
certainty. The greatest care is needed especially when digging up roots, because
otherwise the plant could be destroyed, at least on the site where you found it.
A few plants should always be left behind and you should only pick the part of
the plant that is needed for the treatment.

The Correct Moment

A great many herbs can help us to lead a harmonious and healthy life. In
order to obtain the greatest possible healing power and durability from these
plants, it is important to know about the right moment to gather and store
them.

The healing power of herbs is not evenly distributed over the whole plant.
Many gathering times are very unfavourable, because the active agent is
currently in the flowering herb, while what you may need for your application
is the root. It may be that you are collecting flowers or leaves, while the healing sap is actually gathering new strength in the roots.

When choosing the correct time for gathering, your own instinct and the
state of the weather should always come first. You should always bear in mind,
that the ideal moment for gathering is when the weather is dry. To reap the full
healing powers of herbs you have collected, it is worthwhile waiting for fine
weather. The flower days occurring in Gemini, Libra and Aquarius, are
in general very suitable for picking flowers, but if the sun is not shining and
the weather is cold, then the fact that it's a favourable moment will not help
much. You should always be aware of what is most appropriate at the given
moment.

The best season for gathering herbs is in spring, when the plant is still young
and possesses the greatest healing power. With young plants, the component
substances are released more easily than in older plants, in which they are often
not released at all.

Gathering the Individual Parts of the Plant

It is easier to observe the correct moment to gather the different parts of the plant, than to keep exactly to the correct sign of the zodiac, and this alone can produce good results.

The correct moment to dig up roots is the early spring, when the plant is not yet in full growth, or in the autumn, when the herb has withdrawn into itself once more and the sap has descended again. You should always dig up roots when the moon is full or waning, as they have more power at that time. Roots should not be exposed to sunlight, so the hours before sunrise or in the late evening are the ideal times. The period of the descending moon and the root days of Taurus, Capricorn and Virgo, are also suitable as a gathering time. However, Taurus is not quite so good as the other two signs.

Leaves may be gathered throughout almost the whole year, provided that the plants are young. However, if the plant has been producing sap for a long time, or is already in bloom or has not been mown, it is not very suitable for curative purposes. The best time for picking leaves is late morning, when the dew has evaporated. Leaves should be gathered when the moon is waxing, between the new and full moon. Alternatively, they may be picked when the moon is ascending, (in the period from Sagittarius to Gemini), or on leaf days in the signs of Cancer, Scorpio and Pisces.

Herbs gathered during Scorpio possess a special curative power. In addition, they are particularly suitable for drying, preserving and storing. Herbs gathered during Cancer and Pisces are better used straight away. An exception to this is the stinging nettle. This outstanding blood-purifying remedy should be gathered exclusively when the moon is on the wane, and nettle tea should also only be drunk during the waning moon.

The most favourable times for gathering flowers are spring and summer, when the plants are in full bloom, if possible at midday. Ideally, the sun should be shining, or at least the weather should be warm, so that the flowers are open and the healing force has travelled up to the blooms. Flower gathering should take place when the moon is waxing or full, or alternatively, when the moon is ascending, (from Sagittarius to Gemini), if the weather precludes gathering when the moon is waxing. If you are gathering for winter storage, then the waning and descending moon are also suitable moments, because the flowers will be more certain to dry at that time.

Fruits and seeds should be ripe, neither green nor squashy, when gathered. This usually occurs in summer or autumn. Dry weather is more important than the time of day. However, you should avoid the midday heat. Fruits and seeds gathered when the moon is waxing are only suitable for immediate use. For storing and preserving, the ascending moon (from Sagittarius to Gemini) is a better time. Good gathering days are the fruit days in Aries, Leo and Sagittarius. The most unfavourable days for collecting fruit are those in Capricorn, Pisces, Cancer and Virgo.

THE POSITION OF THE MOON IN THE ZODIAC

The position of the moon in the zodiac has a significant bearing on gathering and using medicinal herbs. A herb that is gathered for the healing or strengthening of those parts of the body that are governed by the zodiac sign of the gathering day, is especially effective. For example, herbs gathered on Virgo days, are especially helpful in dealing with digestive complaints. An excellent foot ointment can be prepared from herbs collected on Pisces days. The following table illustrates these connections.

In the Sign of	Gather herbs for
Aries	headaches, eye complaints
Taurus	sore throat, ear complaints
Gemini	tension in the shoulders, lung complaints
Cancer	bronchitis; stomach, liver and gall-bladder complaints
Leo	heart and circulatory complaints
Virgo	disorders of the digestive organs and pancreas, nervous complaints
Libra	hip complaints, diseases of the kidneys and gall-bladder
Scorpio	diseases of the sexual organs and organs of elimination
Sagittarius	vascular diseases

Capricorn	bone and joint complaints, skin diseases
Aquarius	vascular diseases
Pisces	foot complaints

THE MYSTERIOUS POWER OF THE FULL MOON

The day of the full moon is an outstanding collecting time for almost all herbs and parts of herbs. Roots in particular, when gathered at full moon or when the moon is on the wane, have greater curative powers than at other times. Moreover, roots, especially those which are to serve as a cure for serious illnesses, should under no circumstances be exposed to sunlight.

PRESERVING HERBS

Care is essential when preserving, drying, storing and keeping herbs. It would be a great pity to lose a large quantity of these valuable gifts of nature through neglect. The plants should be put in a shady spot to dry, and be turned regularly. The most suitable underlay is a natural material that is pervious to air (ideally a wooden duckboard, but paper will serve just as well). One should never dry herbs on any kind of plastic sheet. Nothing should be kept for more than a year, which shouldn't be a problem, since fresh supplies of plants are obtainable annually. Don't forget to collect only as much as you need. Moderation, reason and sensitivity should determine the quantity.

The correct time for storing and filling jars or cardboard boxes is always when the moon is on the wane, regardless of the date the herbs were gathered. Never put herbs into containers when the moon is waxing, otherwise there is a danger they might rot. Dark jars and paper bags are the most suitable storage containers. The plants will remain nice and dry, and the aroma and active ingredients will be retained. Plants have different drying times. You must take care to ensure that herbs gathered when the moon is waxing, undergo some of their drying process during the waning moon.

It is not necessary with every type of herb, to dry its individual parts. In the case of many medicinal and culinary herbs, such as marjoram, thyme, lovage and parsley, it is sufficient to hang several plants upside down, like a bunch of flowers, in an airy place, until they have dried. They can be put in containers later in the usual way. This method saves space, looks good, and the aroma creates a pleasant atmosphere in a room. Quick-drying herbs are most suited for this, as then there is no danger that tying them up will cause rotting.

Perhaps, after all of this advice, you will object that not everyone can go herb-gathering in the countryside, or has a herb garden to call his own. Rest assured that plants from a chemist's or herbalist's shop also have their value and often do a lot of good. It is only in the case of chronic, persistent illnesses that the correct moment of gathering is of special importance and ought to be observed at all costs.

PREPARATION AND USE

Often herbs have the best effect when eaten raw as a salad (watercress, young dandelion, and spinach) or alternatively, when cooked (stinging nettles, wood garlic, sage and elderflower). But in addition to this there are many other methods of preparation and use, a few of which we shall discuss here.

✐ Teas and infusions are one of the most common uses for herbs. These are especially suitable for young herbs, containing essential oils, that would evaporate with prolonged boiling. Place as much of the dried or fresh herb as you can hold in three fingers in a cup, pour boiling water on to it, cover and leave to draw for three to ten minutes and then strain (don't use a metal strainer). The tea is ready when the herbs have sunk to the bottom. However, some herbs that contain a lot of oil will not sink, even after several hours. In such cases, ten to fifteen minutes is enough. The tea should be drunk at once, so that it does not lose its medicinal properties.

✐ Boiling (extraction) is suitable for plants containing curative ingredients that are not readily soluble (bitters, tannic acid) and especially for wood, roots or stems. Place the plant matter in a pan, cover with cold water and bring slowly to the boil. The boiling time is a matter of instinct; as a rule not more than fifteen minutes. Some wood and roots need to be boiled for up to half an hour. Where possible, avoid using pots made of steel, iron, copper or brass.

✎ Some herbs cannot withstand either brewing or boiling and should be prepared by cold extraction. Put the plants in cold water and leave them standing overnight. Alternatively, it is sometimes a good idea on the following day to strain the herbs and bring them to the boil in some fresh water (not the liquid produced by cold extraction) and in this way to extract the remaining medicinal properties.

✎ Thicker herbal extracts can be prepared by pouring cold-pressed olive oil over the plant to make mild embrocations.

✎ Many herbs can be crushed to produce a raw juice. The plant juices do not keep and should be used at once (as a drink or perhaps as a compress, depending on the herb).

✎ Tinctures are thin extracts obtained with diluted ethyl alcohol. Place a handful of herbs in dark bottles and add alcohol such as fruit brandy, until the herbs are covered. After storing in a warm place for about two weeks, the tincture will be ready for use.

✎ Plants and plant extracts can also be ground or boiled down with soft fats and made into ointments or mixtures for plasters. Anyone able to buy meat from a farmer who still raises his stock in a natural way, ought to make the most of this and ask him for fat from a pig that was slaughtered at full moon. Perhaps the farmer will know that the meat is much more succulent at that time and keeps better. The fat should be rendered down at a low heat. However, avoid Virgo days, because the meat will easily go mouldy during that period and not keep so well. At home we add the fresh herbs (marigolds are especially suitable) to the heated pork fat and allow it to fry for a short time (as a rule of thumb, about as long as for a cutlet). Two handfuls of herbs to a jam-jar of fat should be enough. Next stand the warm fat together with the herbs in a cool place for about 24 hours. Warm it gently the following day, until the mixture becomes fluid once more. Strain it into clean jars and store in the dark. This ointment is an outstanding remedy for various illnesses, for instance, as a chest embrocation for coughs and bronchitis.

✎ It is important to carry out this work with patience and love and never in a hurry. It is only with care that you can create the right feel for making the preparation and calculate the correct amount of herbs in relation to fat. Always work with enamel pots and use wooden spoons for stirring.

🖉 The correct time for producing ointments is any day between Sagittarius and Gemini – that is to say during the ascending moon. If for reasons of time, you are forced to chose another date, then you should at the very least avoid Cancer and Virgo. An equally good time for making ointments is at full moon. The plants then contain a maximum amount of medicinal properties. After standing for 24 hours, the ointment is poured into jars when the moon is on the wane, which guarantees that it will keep a long time.

🖉 Herb pillows are excellent for aiding relaxation and creating a lovely aroma. They are also very effective in treating rheumatism and allergies. However, do not use protected plants for this purpose. The herbs should be picked when the moon is waxing and then when the moon is waning, they should be put into pillows made of a natural material, such as linen and sewn up. If they are gathered on flower days, their fragrance will give you pleasure for a long time. The choice of herbs depends on the intended use of the pillow. Your herbalist can give you advice and combine the various herbs for you. Even herbs that have been bought should be processed when the moon is waning. Herb pillows can also bring about considerable relief in the treatment of rheumatism and allergies.

🖉 In former times, when there were still enough ferns available, people used to line a whole bed with them. The ferns were sewn in between two sheets and this was used as an under-blanket, especially for people suffering from rheumatism. Likewise, if the patient suffered from cramp at night, he would be advised to fill his pillows with common clubmoss. Nowadays some ferns are protected species, however, it is possible to buy them. Of course the full moon will not have been taken into consideration as the best collecting day; but even so they will still contain curative substances.

THREE PRACTICAL EXAMPLES

🖉 *A remedy to purify the blood* A great many illnesses result in 'bad blood', often recognisable in blemished, unhealthy looking skin, and in higher than normal cholesterol levels. A blood purification cure using stinging nettles, continued until the disorder has receded, can do a lot of good in such cases.

For any healthy individual, a spring cure of nettle tea will drive spring-time tiredness out of his limbs. This cure stimulates the bladder and kidneys, promotes the activity of all the digestive organs and nourishes the body with numerous minerals and vitamins.

When the moon is on the wane (if possible in the afternoon between 3 and 7 p.m.) drink as much nettle tea as you can (about 2 litres; for this quantity use about two heaped tablespoons of the herb). Then wait fourteen days and repeat the cure the next time the moon is on the wane, until the ailment has got better or disappeared. As a cure for healthy people, two periods of fourteen days in the waning moon are sufficient.

Stinging nettles are best collected when the moon is on the wane. Only use young leaves; naturally in the springtime fresh stinging nettles are preferable to dried ones.

If the moon happens to be in an earth sign (Virgo, Taurus and Capricorn) collect a few more nettles than you need for daily use and dry the leaves for winter. However, there is no need for you to carry out a full blood purification cure in winter. On the other hand, it's a good idea to have a nettle tea now and then, after a heavy meal (such as during the calorie-rich Christmas period). In general, earth days are a particularly suitable time to do something for your blood.

The treatment of warts: Warts, moles and strawberry marks should only be removed or treated when the moon is on the wane, regardless of the method you are using. If the treatment has not been successfully concluded by the new moon, stop it completely, and don't start it up again until the next full moon. Treatments during the waxing moon (particularly on Cancer days), can turn out very unfavourably.

The greater celandine is a proven remedy for warts. Begin the treatment on the day of the full moon, spreading fresh celandine sap on the wart every day. The sap is orange-coloured and seeps out of the broken stalk. Take care, as it's poisonous and must not be taken internally. Continue the treatment until the new moon, even if the wart has already disappeared before then.

Garlic can be used during the waning moon to remove those particularly painful warts on the soles of the feet. Cut a hole in a plaster the same size as the wart and stick it on in such a way that the wart remains uncovered. Cut a fresh clove of garlic in half, hold it in place with another plaster over the wart and keep this on throughout the night. In the morning, if possible after taking a shower, remove the plaster. Repeat the process each evening with a new clove of garlic and stop when the new moon arrives. Gradually the wart will go dark and eventually it will be a simple matter to remove it.

Weaning Weaning is very easily done without any medicine, except if it takes place immediately after birth. In the weeks before the full moon, the mother simply puts the baby less and less frequently to her breast, and no longer drinks so much herself. At full moon she breast-feeds the child one last time and drinks very little on this day. A sage tea will further assist in bringing milk production to a standstill.

THE RHYTHM OF THE BODILY ORGANS THROUGHOUT THE DAY

In the course of 24 hours each organ in the body passes through a high-phase, during which it works especially well for two hours and then, as can be seen in the following table, it takes a 'creative break' for two hours.

Every mountaineer knows that if he gets up at 3 a.m., he is off to a much better start than he would be at 5 a.m. At 3 a.m., his lungs work particularly well for a period of two hours. A good start makes it much easier for him to bridge over the drop in energy after five o'clock – by then he will have got into his stride. On the other hand, if he starts at five in the morning, he will have a struggle to get going.

Parents know very well that if they manage to get their children to bed before seven o'clock, they often go to sleep without any problem. But if the 7 p.m. line is overstepped by much, it becomes twice as difficult to lull them to sleep. Between seven and nine in the evening the circulation is working at its best. At this time the body isn't thinking of sleep.

Almost everyone will have noticed that around one o'clock there is often a decline in performance, especially after lunch. The small intestine, which bears the main burden in the digestive process, tries to come into its own. It works especially well at this time and wants the rest of the body to take a break. Its activity is controlled by the vegetative (unconsciously functioning) nervous system, which cannot tolerate any form of stress. Thus we see that the siesta of southern climes finds an echo in the daily rhythms of the organs.

Organ	*High-phase*	*Low-phase*
liver	1 a.m. – 3 a.m.	3 a.m. – 5 a.m.
lungs	3 a.m. – 5 a.m.	5 a.m. – 7 a.m.
large intestine	5 a.m. – 7 a.m.	7 a.m. – 9 a.m.
stomach	7 a.m. – 9 a.m.	9 a.m. – 11 a.m.
spleen and pancreas	9 a.m. – 11 a.m.	11 a.m. – 1 p.m.
heart	11 a.m. – 1 p.m.	1 p.m. – 3 p.m.
small intestine	1 p.m. – 3 p.m.	3 p.m. – 5 p.m.
bladder	3 p.m. – 5 p.m.	5 p.m. – 7 p.m.
kidneys	5 p.m. – 7 p.m.	7 p.m. – 9 p.m.
circulation	7 p.m. – 9 p.m.	9 p.m. – 11 p.m.
general energy accumulation	9 p.m. – 11 p.m.	11 p.m. – 1 a.m.
gall-bladder	11 p.m. – 1 a.m.	1 a.m. – 3 a.m.

If you know the high-phases of the various organs, you will be able to take medicines, flush out toxins or take any other measures to enhance your health at the right moment throughout the day, regardless of the position of the moon. A number of suggestions include a tea to purify the blood, between three and five in the afternoon, forty winks between one and three (a nap in the office), no more breakfasts after nine in the morning, less smoking and drinking between one and five in the morning and so on.

Observe yourself, be aware and make notes. Experience is better than information. Information is only a tool. The hand using the tool needs practice as well. And the same goes for the heart that moves the hand.

THE RELATIONSHIP BETWEEN THE POSITION OF THE MOON IN THE ZODIAC AND BODILY HEALTH

The two to three day sojourn of the moon in each of the twelve signs of the zodiac awakens different forces that are perceptible everywhere in the animate world and have a marked effect on our bodies. The principle, starting point and effect of these forces are not especially difficult to describe; but on top of that they possess a kind of 'colouring' which isn't easy to put into words – something that leaves its mark on our mental and emotional mood, like a musical chord that sounds from afar and can be heard by those ready to listen. On the following pages, you will find a summary of the individual impulses and their implications for health, ordered according to zodiac sign. You can use the information concerning the forces prevailing in the different signs of the zodiac to take precautionary measures to protect your health.

♈

ARIES

In the waxing moon
October to April
In the waning moon
April to October

Colouring: Aries is energetic, at times impatient, 'running its head against a brick wall'. At this time unseen chains rattle audibly and are less willingly borne. Things get going and the way straight ahead seems the best.

Aries influences the head region. Anyone who is especially susceptible to migraine will often suffer badly during the two or three Aries days in the lunar month. I have often been able to observe that many people, especially women, are plagued with violent headaches on these days. Frequently, they themselves contribute towards the outbreak of headaches through their behaviour prior to the Aries days. They have a talent for putting off important matters, household chores and appointments shortly before the Aries days, until precisely on the Aries day itself it all comes crashing down on them. Observe yourself and try to keep the Aries days as free of stress as you possibly can.

A useful measure for avoiding migraine, is to drink plenty of water on Aries days and go without coffee, chocolate and sugar. Most importantly, listen to your body's signals and to what agrees or disagrees with you.

The sign of Aries also affects the eyes and the brain. Eye compresses for inflamed or exhausted eyes applied on Aries days are bound to be effective. Herbs that are known to treat headaches and eye complaints develop greater strength when gathered during Aries. Injuries to the eyes occurring in Aries are more harmful than at other times. Be careful to avoid straining your eyes during this susceptible period.

Especially critical Aries days occur in March, April, September and October. Everyone who suffers from frequent headaches should arrange for these days to be as peaceful as possible. All operations on the head should be avoided. The Aries days in October are especially bad because they fall directly at full moon.

TAURUS

In the waxing moon
November to May
In the waning moon
May to November

Colouring: Realism is the order of the day, material security is a virtue. Persistence is easier, thoughts and reactions slower. Obstinacy prevails.

As the moon enters Taurus, the vocal cords, the jaws, teeth, tonsils, the thyroid gland, the neck and ears are all affected. The following pieces of anecdotal evidence illustrate this point.

For weeks a young man goes riding in his convertible in beautiful, warm weather and enjoys the fresh air. Then one day, quite literally out of a clear blue sky, he gets a stiff neck, is reduced to giving himself neck compresses all day long and feels like an old man.

Or you suddenly develop a sore throat and notice at the same time that friends, neighbours and colleagues are going around with croaking voices and scarves about their necks, even though colds and throat inflammations are not always infectious. In many cases this is Taurus 'kicking in'. Which doesn't mean of course that one is bound to get a sore throat at this time. But there is definitely a greater danger of developing a sore throat during Taurus.

Anyone who has taken a tea for hoarseness or inflamed tonsils on Taurus days, knows just how effective it can be. Other cures for throat inflammations are also especially effective during these days.

Giving a speech on Taurus days can be agony for the inexperienced and may end in hoarse croaking. On Taurus days, when the weather is cold, you should always wear a hat to protect your ears, because one's ears are more sensitive to draughts at this time. A drop of Saint John's wort oil put in the ear, now and then on Taurus days can often prevent earache, especially when the Saint John's wort flowers used to make the oil were gathered during Taurus.

♊	*In the waxing moon*
	December to June
	In the waning moon
GEMINI	June to December

Colouring: The mind becomes active and versatile, moving in leaps and bounds. A breath of wind can deflect it from its course. The forces branch out and penetrate into every corner.

The moon in Gemini influences the shoulder area and the lungs. Rheumatic gout in the shoulders responds particularly well at this time to suitable ointments, produced preferably from herbs that were gathered during Gemini and Taurus. Gemini days are ideal for working on the shoulder area. Some properly thought out exercises can work wonders – a treat for your shoulders. However, you will not necessarily be spared stiff muscles afterwards. In any case, this is probably a good sign, for that is how the body signals that it is busy with detoxification.

Work on the lungs at this time is also very beneficial. For instance, practising some breathing exercises can be extremely helpful.

In the waxing moon
January to July
In the waning moon
July to January

CANCER

Colouring: Feelings gain depth, but also weight. The inner becomes more colourful than the outer. It becomes easier to make sacrifices. Dense, extensive growth takes place. The ground shakes. This is often a period of slight restlessness.

During Cancer, the chest, liver and stomach are particularly susceptible to lunar forces. For instance, you are more likely to suffer heart burn and wind and should eat less during this period. Staying up all night at this time will put excess strain on your liver and you will feel totally shattered the next day. If you happen to be susceptible to complaints of the liver, gall-bladder, lungs or chest, you can take advantage of Cancer days to do these organs some good.

From July to the following January, Cancer days always occur in the waning moon, and for the next six months in the waxing moon. You will recall that when the moon is on the wane, the system is cleaned out and that when it is waxing, it is supplied with as much goodness as possible. This means that in the case of the stomach and liver, healing or expelling poisons has more chance of success in the period from summer to winter, than from winter to summer.

Anyone who suffers from rheumatism should not hang bedding over the window-ledge or balcony to air during Cancer (a water sign). The damp remains in the feathers and one feels chilly all night.

In the waxing moon
February to August
In the waning moon
LEO   August to February

Colouring: Muscles swell, determination reigns, people take heart. Limits lose their sharp outlines and appear to be more easily surmountable. Pleasure in risk; fire that dries out.

The lunar rhythms during Leo influence the heart and the circulation. On Leo days, the blood circulation is working at its most efficient. However, anything that could strain the heart and circulation should be avoided if possible. You may be more susceptible to backache at this time and to heart trouble. Sleepless nights can cause quite a lot of problems during Leo. However, by the time Virgo comes, everything settles down.

People with heart problems can sometimes already sense during the sign of Cancer that Leo is on the way. Heart sufferers should steer clear of strenuous journeys or undertakings at this time. Many an inexperienced mountain walker puffs his way to the summit during Leo, even though he usually has no problems with his circulation or heart.

Leo days are very good for collecting herbs that have a curative effect on the heart and circulation. Although the sign of Virgo is responsible for the digestive organs, anyone thinking of taking curative or tonic measures ought to start in Leo.

In the waxing moon
March to September
In the waning moon
VIRGO   September to March

Colouring: The logically arguable end justifies the means. Small, hesitant, methodical steps, almost pedantic. First investigate, then act. Things are separated and split up with the object of developing them.

During Virgo, the moon's influence on the digestive system is particularly strong. Sensitive people in particular, often have problems with their digestion at this time. It is best to eat as healthily as possible during these days and to avoid heavy or fatty meals.

Herbs gathered during Virgo have a favourable effect not only on the stomach, but also on the blood, nerves and pancreas. In particular a blood-purifying infusion, such as stinging nettles gathered in Virgo, is bound to be beneficial, especially for an enlarged pancreas.

The winter store should not be laid down until September, when Virgo appears once more in the waning moon.

	In the waxing moon
♎	April to October
	In the waning moon
LIBRA	October to April

Colouring: The artistic rules, but so does indecisiveness. Tactful sensitivity, without much punch. Swinging this way and that, until equilibrium is achieved.

During Libra, the moon exercises a strong influence on the hip region, the bladder and the kidneys. One is more susceptible to bladder or kidney inflammations at this time. Take special care to keep the area of the bladder and kidneys well insulated. Sitting on stones or wet grass in Libra is asking for trouble.

A useful measure during Libra is to drink a great deal between three and five in the afternoon, in order to rinse the bladder and kidneys thoroughly.

Special exercises for the hip region are particularly helpful at this time.

Often I am asked the best time, in terms of the position of the moon, to carry out a hip operation. The correct moment for this is when the moon is waning. In the months from April to October, the star sign of Libra never occurs in the waning moon. In addition, the star sign that rules the part of the body under consideration should never be the current one in the calendar. If you only have time for an operation between October and April, then avoid the star sign Libra in the waning moon.

SCORPIO

In the waxing moon
May to November
In the waning moon
November to May

Colouring: Opportunities are precious, and time is a sharp sword. Acceptance is difficult. Energy goes deeper, bores its way down and searches. Darkness beckons.

No zodiac sign has as strong an effect on the sexual organs as Scorpio.

As a precaution, hip baths using yarrow can be of assistance at this time, in a great many female disorders. Mothers-to-be should beware of any exertion on Scorpio days, because miscarriages can happen more easily especially when the moon is waxing. The ureter, too, is especially sensitive during Scorpio and will respond well to tonic treatments. During Scorpio, it is advisable to keep your feet warm and the region of the pelvis and kidneys too, to avoid kidney and bladder inflammations. Anyone suffering from rheumatism ought not to air bedding outside on the window ledge or balcony during Scorpio days, as being a water sign, moisture will remain in the feathers.

All medicinal herbs, sown or gathered during Scorpio, are especially effective. Fetch all your herbs home by May or June to make a pillow, then they will give you pleasure for many years to come. However, do not fill the medicinal pillows during Scorpio, because a water sign is not very suitable on account of the moisture.

SAGITTARIUS

In the waxing moon
June to December
In the waning moon
December to June

Colouring: The future seems more important than the present and past; the great more important than the small; bringing together more important than splitting asunder. Magnanimity reigns, but so, too, does pathos. The stride lengthens.

The influence of the moon in Sagittarius affects the hips, pelvis, ilium, sacrum, and thigh-bone. The sciatic nerve, veins and thighs make their presence strongly felt on Sagittarius days. Massages at this time can do a lot of good and loosen up tense muscles. On the other hand, a body that is out of training will feel the effect of long hikes on the thighs particularly strongly during Sagittarius. So one shouldn't overdo it during Sagittarius and go on long hikes without some initial training. Anyone who picks on Sagittarius to take the kids on their first major mountain walk, or perhaps even forces them into it, could very well put them off hiking for a long time to come.

In the waxing moon
July to January
In the waning moon

CAPRICORN   January to July

Colouring: The air becomes transparent, thinking clear and serious, straightforward, somewhat inflexible. The goal seems more important than the way there.

The moon in Capricorn has a powerful influence on the bones, joints, knees and skin.

Any excessive strain on the skeleton in general, and the knees, in particular, can have serious consequences during Capricorn. Just as in Sagittarius, if you are a beginner, or are out of practice, you should not start out on mountaineering or skiing tours at this time. Surgeons and orthopaedic specialists know precisely when the moon is passing through Capricorn as the demand for knee operations increases. Footballers with knee problems should under no circumstances overdo it at this time.

Knee poultices are particularly useful, either as a precaution, or a cure during these two or three days. Other bones and joints will also benefit from treatment during Capricorn. These days are also very suitable for any kind of skin care.

AQUARIUS

In the waxing moon
August to February
In the waning moon
February to August

Colouring: The mind cuts capers, intuitive thoughts get a hearing. Shackles are not tolerated – not even imaginary ones.

In Aquarius, the moon's influence is felt specifically in the lower legs and ankle joints. Inflammation of the veins is not uncommon during Aquarius. Now is the time to put some ointment on your legs and rest them in a raised position. Anyone who is inclined to get varicose veins should avoid long periods of standing on these days. Operations on varicose veins should not be carried out during Aquarius. Even a longish stroll around town can be a nightmare on Aquarius days. Taxi drivers generally have more to do at this time.

PISCES

In the waxing moon
September to March
In the waning moon
March to September

Colouring: The common good comes before individual advantage. Borders become blurred, it is easier to take a look behind the scenes, and the world is tinged with fantasy. Rigid points of view are avoided; plenty of bowing and scraping occurs.

The moon in Pisces influences the feet and toes. Pisces is the best time for foot-baths and the treatment (but not the removal) of corns. During Pisces days, warts on the feet can be treated with considerable success. However, it is essential to look out for the waning moon. If the moon is waxing, it may well be that after a treatment you suddenly have five warts instead of three.

On Pisces days, everything that you allow into your body, alcohol, nicotine, coffee, and medicine, has a much stronger effect than at other times. Presumably this is because the meridians of all the internal organs terminate in the feet, and thus react with especial sensitivity on Pisces days.

Lunar Rhythms in the Garden and the Countryside

A man is born gentle and weak.

At his death he is hard and stiff.

Green plants are tender and filled with sap.

At their death they are withered and dry.

Therefore the stiff and unbending is the disciple of death.

The gentle and yielding is the disciple of life.

Thus an army without flexibility never wins a battle.

A tree that is unbending is easily broken.

The hard and strong will fall.

The soft and weak will overcome.

LAO TSU

There are many reasons in gardening, agriculture and forestry for returning to the observance of the phases of the moon and the position of the moon in the zodiac. One of the most important is that with the aid of lunar rhythms, it is possible for us to move away from an overdependence on pesticides, insecticides and fertilisers and find our way back to a natural and dynamic balance in nature. The gardening and agriculture of the future will have no other choice, because one cannot exploit nature indefinitely. So why shouldn't we start straight away?

A short while ago I was flying from Hamburg to Munich. The weather was perfect and I had a window seat. I couldn't keep my eyes off the ground ten kilometres below me. It seemed to me there wasn't a single spot that wasn't utilised, cultivated, built up and furrowed with roads. Even if I saw a largish, continuous stretch of woodland, there would be some gravel works or other building sunk right in the middle of it. The thought came to me how lovely it would be if we could learn once more to treat nature in such a way that this view would only awaken pleasant feelings.

With totally unpredictable results for the entire cycle of nature, millions are spent today on research into breeding and genetic engineering, with the purpose of transforming plants in such a way that, on command, they will do what they would have done naturally, given the correct choice of moment for planting, nurturing and gathering.

This situation is reminiscent of the 'good intentions' of nutritional scientists at the turn of the century. They observed that in the process of eating, certain substances enter the body and are subsequently expelled again unchanged. They concluded that these substances were superfluous and began to rid foodstuffs of them, and produce ready-made foods. The consequences are well-known.

The Carrier Parrot

A man crossed a carrier pigeon with a parrot, so that their descendants would be able to speak the message instead of having to carry it on a piece of paper.

But the bird that resulted from this experiment took hours to complete a flight that normally would only have taken minutes.

"What kept you?" asked the man.

"Well," said the bird, "it was such a nice day that I decided to go on foot."

The Price of Scientific Advances

Advances in science and technology have convinced us that practically all problems can be solved through their agency, even problems that we would never have encountered if we had not relied so heavily upon science in the first place.

The initial successes were really enormous – crop yields increased, pests disappeared and unlimited opportunities seemed to open up. Only gradually were people forced to recognise that exploitation of nature is not the ultimate wisdom.

This book is intended to help wisdom and reason come into their own once more in the sphere of gardening, agriculture and the countryside in general. It can offer inspiration to those prepared to consider the advantages of a return to natural means of production. You will find many suggestions on the following pages that will make it easy to give up using chemicals and pesticides and to rely on nature's rhythms. It is possible to garden and farm without using poisons and without the nonsensical manipulations of genetic engineering and to produce equally large yields and a harvest of far superior quality.

There are so many proofs of the benefits of working with nature: fruits that only carry within them the strength of the sun, which gave them their colour, cereals that let us taste the meaning of harmony between heaven and earth, vegetables that carry this harmony into our bodies, earth that willingly bestows these gifts on us for centuries on end, without the use of fertilisers and other poisons.

Perhaps you have one or two experiences of your own that come to your aid. Almost everyone who has anything to do with gardening and country life has experienced things that on closer inspection simply cannot be explained. At various times, under completely identical conditions, we sow, plant, water, transplant, fertilise, harvest and store, but with totally different results. Sometimes the lettuce forms a nice head, sometimes it sprouts flowers and seeds and cannot be used. Sometimes a farmer harvests the most wonderful turnips, while his neighbour, using the same seeds, and with the same weather conditions and soil, is dissatisfied with the result. Then again, one's own crop is attacked by pests, whilst the neighbour's remains untouched. Or a small strip at the edge of a cornfield flourishes much better than the remaining area. Sometimes the part of the potato plant above ground grows splendidly, but the crop itself remains undeveloped. At other times, the plant looks stunted, and yet it yields the most enormous potatoes. Sometimes the cherry jam lasts for years, while at other times jam made of the finest fruit can quickly go mouldy, even though everything was preserved under the same conditions.

Frequently, we blame the weather or the quality of the seeds on bad results. In many cases that may well be true, but much more frequently the reason is simply that attention was not paid to the correct timing of the task in question.

You may have noticed already that in a lettuce or a cabbage plot, all the plants sprout or all the plants form a head. You never find both conditions simultaneously in the same bed. It is almost never the seed that is at fault.

Simply giving up spraying and using fertiliser and just following the rhythms of the moon, will not automatically produce fantastic results. Because our soils are heavily contaminated, it may take several years of patient work to create healthy soil. If you are prepared to proceed slowly, the knowledge which people have been testing and using for thousands of years before you will be your reward. None of it is new. It is merely that technical progress has seduced us into supposing that we can afford to push this valuable inheritance into the background, or even let it sink entirely into oblivion. Have courage, it doesn't cost anything, and in any case the work on the fruit, vegetables and cereals has to be done. While you are sowing, planting or harvesting, pay attention to the signs of the zodiac and the phases of the moon. In due course you will be pleasantly surprised by your success, in every respect.

Do not despair if deadlines or weather conditions spoil your plans and prevent you from observing the correct times for tasks. You will discover that there are several alternatives for every job in the garden or the field that are almost as suitable. However, always avoid the unfavourable times.

Of course we cannot discuss every plant in detail here: that would be missing the point. But the principles can easily be applied to other tasks in the garden and the field.

My reason for writing this knowledge down is quite simple. It is so that you may obtain, if that is your desire, a knowledge that can accompany you throughout your whole life, without having to be forever looking things up in textbooks and guides. A knowledge that becomes second nature to you. Even if, unlike me, you did not grow up with this knowledge, you now have the opportunity to accumulate your own experiences, and that is a thousand times more useful than examples in a book. After you have tried it out a few times, you will soon notice just how simple everything is.

Perhaps my personal outlook will be of further assistance to you in appreciating the advantages of a thorough rethink. I also enjoyed my early days in

Munich to the full, and very quickly forgot my responsibility towards the natural world, because of course, in cities, you can buy anything you want. Prices seemed so low to me in comparison with income. It is all too easy to get the impression that physical labour is no longer worthwhile, since fruit and vegetables are relatively cheap. However, when I came back home after quite a long time and ate the local lettuce, I noticed an enormous and unmistakable difference. But at that time this was not yet sufficient reason for me to return to natural ways. I had to fall ill first. Even then I tried everything else before I fully realised that I couldn't and wouldn't go on living like this. I learnt once more to take responsibility for my own body, and understood that I can only eat good vegetables if I actually have the possibility to get hold of healthy vegetables. The wheel had turned full circle and I accepted the years of thoughtlessness as a learning process. So all that is needed is experience and clarification. That is why it is not difficult either for a town dweller to realise after a certain amount of observation that a short-term success is no success at all. It is essential for us to learn again how to accept responsibility and not let ourselves be blinded out of sheer laziness.

Every sign of the zodiac has an effect on a specific part of a plant, on either its roots, leaves, fruit or flowers. The table at the end of Chapter I summarises the most important impulse characteristics of the zodiac signs and how they influence all work in the garden, on the land and in the countryside, including their effect on the different parts of plants, the day quality and the ascending or descending force inherent in each sign. By referring to this table and the calendar included at the back of the book, you will be able to interpret the tips that are given in the following chapter and adapt your gardening work to these rhythms. By taking into account the most favourable times for the various tasks in the garden, you will avoid many negative influences and your successes will be a source of pleasure for you.

Sowing and Planting

The main work both in the garden and in the field, generally begins in the spring with turning over the soil, followed by sowing and planting. The correct timing of these tasks is extremely influential on the growth and ripening of plants and their resistance to weeds and pests.

In Chapter II we discussed how greatly our bodies and health are affected by the varying impulses of the waxing and waning moon, the full moon and the new moon and the position of the moon in the zodiac. In addition to that, in the plant kingdom the varying forces of the ascending and descending moon also have a part to play and can be used in many ways. For instance, they can often be employed as an alternative when the most favourable time for a particular task in the garden or the field has to be missed.

An important time for a great many planting activities is the approximately thirteen-day period of the descending moon, between Gemini and Sagitarrius.

In the following pages, when we talk of the descending moon, you must not forget that this has nothing to do with the waning moon. However, if you look closely at the calendar at the back of the book, you will see that both rhythms can overlap and influence each other.

THE WAXING OR WANING MOON?

When the moon is waning, our bodies are tuned to giving, expending energy and activity; whilst when it is waxing they are tuned to breathing in, planning, conservation and gathering strength. With the soil the exact reverse is the case. When the moon is on the wane, the sap runs to the roots; the soil is receptive and breathes in, whilst when the moon is waxing, the sap climbs higher and growth above ground and breathing out predominate.

Before we come to the rules for sowing and planting, a brief tip about digging over a newly laid-out garden plot in spring might be useful, since this always precedes the planting work.

In the spring, dig over each bed three times. The first time, when the moon is waxing in Leo; then in Capricorn with the moon on the wane, and finally a third time, preferably when the moon is waning.

In spring, appropriately enough, the zodiac sign Leo always occurs when the moon is waxing, and Capricorn when it is waning. Weeding and turning over the soil, with the waxing moon in Leo, stimulates the growth of the seeds of the weeds buried in the soil and everything sprouts and germinates. The same action, with the moon on the wane in Capricorn, ensures that the weeds disappear and almost none return because there are no more seeds in the ground. If you keep to this rule for turning over the soil, you will have created the best conditions for profiting from the tips that now follow.

The basic rule for planting and sowing is as follows. Plants and vegetables that grow and flourish above ground should be sown when the moon is waxing, or alternatively, when it is descending. However, the rules for leaf vegetables are different. These should be sown and planted when the moon is waning, preferably during Cancer. Vegetables that grow below ground will thrive if the waning moon is taken into account when selecting the sowing or planting day. If this time is not practicable, then an alternative time can be chosen when the moon is descending.

With the aid of the calendar at the back of the book you will have no difficulty in selecting these appropriate moon phases and at the same time paying attention to the correct zodiac sign.

THE CHOICE OF ZODIAC SIGN

When it comes to choosing the correct sign of the zodiac for planting, this depends on what type of plant you are growing. Tomatoes, for example, are a fruit and should be planted on a fruit day in Aries, Leo or Sagittarius. Leaf vegetables, such as spinach and lettuce are best planted or sown on a leaf day in the calendar, in Cancer, Scorpio or Pisces. When planting or sowing lettuce, however, there should be also always a waning moon.

The same principle applies to root vegetables. For example, celeriac, carrots, onions and radish should be planted on a root day in Virgo, Taurus, or Capricorn. An exception to this rule is the potato. Although the waning moon is the correct time for planting, you should not plant too close to the new moon, but shortly after the full moon. For flowers and most medicinal herbs, a flower day is best, in Gemini, Libra, or Aquarius.

When you have grasped these principles you will be able to create a garden planner for the whole year. Of course time and weather constraints may not always allow one to hit exactly the right day. However, there is plenty of leeway, as you will discover.

WATERING AND IRRIGATION

On the subject of watering and irrigation, here is some advice which may seem rather provocative to many a passionate gardener. Once seeds are in the ground, it is quite sufficient to water them once at the outset. If there happens to be a drought at the time, you can continue watering for a few more days, but for no longer than that. In our latitudes, additional watering is utterly pointless.

Nowadays the soil of many gardens and fields is regularly watered, regardless of the natural conditions. This spoils the earth and makes all the plants lazy and listless. The roots grow shallow and no longer reach deep into the ground, fertiliser is washed away and all the goodness is taken out of the produce. In the natural rhythm of rain and drought, both soil and plant wake up, have a good stretch and begin to breathe. A plant left to draw water from the earth knows that every drop counts, and it catches up on what it needs. Its inner strength is quite different, and so is that of its fruit. However, it is not advisable to stop watering plants suddenly without some preparation. The soil has to adapt gradually to natural rhythms. Just like a muscle that has become flabby through misuse, first there comes the training, then the aching stiffness and then the strength.

House plants and balcony plants, on the other hand, need to be watered, though not as often as is frequently done. House plants preferably should be watered on leaf days in Cancer, Scorpio, or Pisces, ideally with calcium-free rainwater. Perhaps this piece of advice will astonish you or even strike you as cruel, for leaf days only turn up at intervals of six to eight days. And yet watering only on these days is quite sufficient (with the exception of some exotic plants). Plants that have a high water requirement simply should be watered several times a day, possibly on all two or three of the leaf days.

Even if I go away on a two-week journey, I don't need anyone to come round and water my house plants for me. If I water them generously for the last time on a water day, so that if possible there's still some water in the saucer, all my plants will hold out. You should accustom your plants slowly, not abruptly, to this new rhythm. Exceptions to this are some very thirsty house plants and garden plants, such as tomatoes, which need more frequent watering.

At the very least you should give up watering on flower days. Pests often spread on plants that are watered on flower days in Gemini, Libra, and Aquarius, lice in particular. Putting house plants out in the open, with the good intention of exposing them to the rain, can have equally unfavourable effects, since the leaves are often unable to withstand direct watering.

FRUIT SEQUENCES AND PLANT COMMUNITIES

When growing vegetables, the order in which you plant each vegetable, known as the fruit sequence, the alternation of one species with another and the choice of which plant is grown next to another, are of particular importance. All gardeners will know about this and there is information about it in many gardening books – plants that mutually foster one another and protect each other from pests, unsuitable plant communities and much more. However, there are one or two suggestions for beginners here.

One particularly favourable fruit sequence is one in which vegetables and arable crops are cultivated that grow above and below ground level in annual alternation. Within the vegetable plot itself, organic gardeners take care to place plants with shallow roots next to plants whose roots go deep. Since the harvest times are different, the plant with the longest ripening period ends up having the most room, because the other vegetables in the meantime have been harvested.

Particularly favourable plant communities

carrots next to onions	lettuces next to radish
tomatoes next to onions	peas next to celery
tomatoes next to parsley	potatoes next to cabbage types

Favourable communities

cucumbers go well with: onions, runner beans, celery, beetroot, parsley, lettuce, kohlrabi, cabbage varieties, dwarf beans

potatoes go well with: spinach, dwarf beans, kohlrabi, dill

celery goes well with: dwarf beans, spinach, onions, runner beans, tomatoes, leeks, kohlrabi, cabbage varieties, cucumbers

parsley goes well with: tomatoes, onions, radishes, cucumbers

tomatoes go well with: celery, spinach, onions, parsley, cabbage types, kohlrabi, lettuce, leeks, dwarf beans, carrots

spinach goes well with: tomatoes, runner beans, strawberries, kohlrabi, carrots, potatoes, cabbage varieties

lettuce goes well with: onions, tomatoes, runner beans, dwarf beans, radishes, dill, peas, cucumbers, strawberries, carrots, cabbage varieties, leeks

onions go well with: tomatoes, strawberries, cucumbers, parsley, lettuce, kohlrabi

strawberries go well with: carrots, leeks, cabbage varieties, radish, lettuce, spinach, onions

Particularly unfavourable plant communities

beans next to onions

cabbage next to onions

potatoes next to onions

red cabbage next to tomatoes

parsley next to lettuce

beetroot next to tomatoes

tomatoes next to peas

peas next to beans

TRANSPLANTING, RE-POTTING AND CUTTINGS

Perhaps this list of compatible plant communities may inspire you to put some of these plants in your garden. If you are moving plants from one area to another, there is a correct time to do this as well. Transplanting should take place when the moon is waxing, or alternatively when the moon is descending (from Gemini to Sagittarius).

Plants that are moved to another place or another pot at this time grow new roots quickly and thrive marvellously. With older plants, and above all with older trees, it is important to pay close attention to the time of transplanting. The correct moment for transplanting older plants and trees is in spring or autumn, at the time of the descending moon, during Virgo. At that time, even a fairly old plant or tree will put down roots again. For cuttings, too, the period of the waxing and descending moon is favourable. They settle quickly and in a short time develop new root-hairs. Again, the Virgo days are the most suitable. However, when planting cuttings in autumn, you should look out for the waning moon.

Prevention is better than cure

The method of fruit sequence described above, is a good preventive measure against a serious infestation of pests.

The table at the end of Chapter I shows how every sign of the zodiac has an effect on a particular part of a plant (Aries on fruit, Taurus on roots, etc.). If, for instance, you repeatedly tend or water a vegetable patch when the influences of the zodiac are unfavourable, you are setting up a breeding ground for unwanted vermin.

The best way of preventing an infestation of pests is to plant and sow at the right moment, taking into account the influence of leaf, fruit, flower and root days on flowers and plants. Sometimes the weather can wreck these calculations; but at the very least planting, sowing and tending should not take place on a day that is actually unfavourable.

Fruit – Aries, Leo, Sagittarius   **Flower** – Gemini, Libra, Aquarius

Root – Taurus, Virgo, Capricorn   **Leaf** – Cancer, Scorpio, Pisces

Combating Weeds and Pests

Here are two brief pieces of information on this extremely important subject (both from the *Süddeutsche Zeitung*, 25.4.1991) in order to bring home to you the full importance of the problem.

"One kilogram of triazine, a herbicide that is sprayed in incredible amounts (recently banned, but even the substitute materials are not exactly suitable as a sauce for potato pancakes), costs about DM 60 (£30) in the shops. In order to remove the same amount of triazine from our groundwater requires 1000 kg activated carbon costing DM 10,000 (£5,000) not counting the cost of removing the contaminated carbon, which also has to be rendered harmless."

"In 1940 farmers used very little insecticide. At that time pests destroyed about 3.5% of the harvest. Today a thousand times as much pesticide is sprayed. One would have thought that this quantity would be quite enough to finish off the very last cabbage white. Wrong: crop losses have currently climbed to 12%."

If every manufacturer of time bombs such as this had to bear the cost of cleansing the environment of his poison, then the world would look very different. The old wisdom about natural rhythms would not have been lost, since the necessity for applying it would have continued to be felt. The tips that follow for avoiding and getting rid of pests and weeds will cost you nothing other than a little patience.

We have already mentioned that many plants designated as 'weeds' – dandelions, stinging nettles, daisies, celandines and many others, are in almost all their parts valuable medicinal herbs for the widest variety of ailments. At the same time, when they decompose, their great powers contribute towards the re-establishment of biological balance in a soil that has been exhausted.

In the same way, pests are often beneficial. Perhaps not for us, if we measure everything by the yield of the harvest or the 'beauty' of the harvested fruit. But for a large number of animals – birds, beetles, caterpillars, rodents, and many other creatures great and small, they provide essential nourishment. Each of these animals is a link in an endless chain, a spiral slowly turning into the future, to which we have given the names 'nature' and 'evolution'. Certainly nature can dispense with one or other species of plant or animal, as it has done many times in the past and allow it to die out. But we humans cannot do so. With every exterminated species of animal and plant, there dies a piece of ourselves, of every single one of us. Until finally nature will dispense with us.

In spite of all this, so many garden lovers go into panic at the sight of a dandelion. They dash into the shed and rummage around for the chemical bludgeon. It is this attitude that has contributed to the fact that the earth in private gardens and allotments is many times more poisoned than the ground used for agricultural purposes, even for monoculture farming. Of the 30,000 tonnes of pesticide in 1724 different products with 295 poisonous ingredients, which land every year on German soil alone (and finally end up in the earth, in the groundwater, in our muscles, skin and entrails) 2000 tonnes are sold to small-time gardeners and gardening enthusiasts, mostly in order to keep lawns in decent condition.

However, anyone whose wish is not to eradicate nature, but to live in harmony with it, ought first and foremost on sighting pests to ask himself the question, "Are they really pests?"

And when, using moderation and common sense, you have answered the question as to the nature of the pest and have come to the conclusion that you want to do something about it, it is then time for a second question. "What is the cause of the infestation?"

In the answer to this is often concealed the appropriate measure needed to get rid of the nuisance, or at least avoid its reappearance the next year. To be sure, there are many possible reasons for a massive attack of pests, and it is certainly not easy to determine the exact cause. You may ask yourself, did I make a mistake in cultivation and care? Is it possible that I chose unsuitable soil?

The answers to these two questions can help us a great deal.

MIXED CULTIVATION

As every farmer and gardener knows, planting compatible plants together in the same plot can be very effective in keeping pests away at the very outset. This method is known as mixed cultivation. It is an immense advantage if plants can help one another to keep the pests in check.

Here are some suggestions for suitable plants that can be used as remedies to deal with the most familiar garden pests. You should put the plants in the area where the pests are prolific. When using these remedies, you should take care to plant all herbs when the moon is waxing and all bulbs when the moon is on the wane.

However, sometimes the only remedy that appears to work is to prune the plant radically. It is essential that this is done with the waning moon in the fourth quarter or ideally, exactly at full moon. In most cases, the plant will then recover.

Against	*helpful*
cabbage whites	peppermint, sage, tomatoes, thyme, mugwort
aphids	ladybirds, nasturtium (esp. under fruit trees), extract of stinging nettle
mites	raspberries
sawflies	tansy
flea beetles	elder-flower extract, wormwood, peppermint, onions, garlic, lettuce
ants	lavender, lamb's lettuce, tansy, dead fish (buried)
mice	garlic, crown imperial, ribbed melilot
mildew	garlic, chives, basil
carrot fly	onions, sage
fungus	chives, field horsetail
mould	bulbs
moles	when the moon is waxing, open up the hill and expose the hole, either by hand or using a harrow

When using extracts to combat pests, it is best to proceed as follows. Before full moon, place two large handfuls of the plant in question in 10 litres of cold water. Allow the extract to stand for twenty-four hours and then pour it undiluted onto the root area and into the soil around the trunk of the infested plant (not on the trunk itself, or the stem, stalk, leaves or flowers). If you are applying the extract when the moon is waning, you should allow it to stand for twice as long. Do not throw the remaining extract away. After diluting it, it will serve as a good fertiliser for days.

If you have taken planting and tending times into account and yet pests still occur in large numbers, then there are some more tips you can use concerning the phases of the moon to combat pests.

With one or two exceptions, the following may serve as a rule of thumb. For all measures to combat vermin, the waning moon is appropriate. Vermin that live in the soil are best dealt with on a root day in Taurus, Virgo, or Capricorn.

Combating pests that live above ground is especially effective when the moon is in Cancer, but Gemini and Sagittarius are also suitable times.

The rules for combating snails and slugs in the garden and on the land are different. The waxing moon in Scorpio is the best time for this. Fortunately, nature has so arranged things that Scorpio generally occurs during the waxing phases of the moon in spring, precisely when slugs and snails are sounding the attack.

Collect as many eggshells as you possibly can (the shells of boiled eggs are not suitable) and crunch them up into small fragments when the moon is on the wane. Gather up the snails that are in the plot. Then, when the moon is waxing, scatter the fragments around the plants and all over the plot. The sharp edges of the shell fragments act as an effective deterrent to the delicately skinned creatures. If the infestation is particularly bad, repeat the remedy the following month when the moon is in Scorpio.

It is essential that the moon is waxing when the shells are scattered, because otherwise the next rainfall will wash them into the soil. When the moon is waxing, the earth does not absorb so much moisture and solid material stays lying on the surface. After some time the shells disappear into the ground (which can have the useful side-effect of making the soil more limy); but by then the danger of infestation from snails normally has passed.

Other suitable means for dealing with slugs and snails are spreading wood ash and sawdust over the soil or growing a combination of garlic, sage and nasturtium, to name but a few examples.

Furthermore, slugs are a real treat for some of their natural enemies, in particular toads and frogs. But these animals are only at home in gardens that are free of poison. If the environment is conducive, say for instance they are attracted by a small pond, then they will come along of their own accord, even in the town. On a farm, a water meadow or a little stream is often enough to attract them.

If you wish to introduce frogs into your habitat unnaturally, pay close attention to the sign of the zodiac and the day of the week. Whether the moon is waxing or waning is not of great importance, but if you can choose, set them loose when the moon is on the wane. The choice of the correct zodiac sign is especially important. Do not do this in Cancer, Leo, Taurus or Aries. The frogs will not feel at home and will disappear again after some time, or even die. The other signs are more neutral and more suitable for such purposes. Frogs should not be set loose on a Tuesday or a Thursday. These days are unsuitable for releasing animals of any sort (for instance, after buying them, or after moving house).

The hedgehog, too, is a natural enemy of slugs and snails and eats them in large quantities. A natural garden with an autumn heap of brushwood is an enticing proposition for our prickly friend. The best time to pile up the brushwood is when the moon is waning, then it will stay nice and dry. Likewise, autumn leaves raked together in a pile can help the hedgehog to survive the winter without coming to harm. The leaves should also be raked together when the moon is on the wane and preferably during a dry sign (not in Cancer, Scorpio or Pisces).

COMBATING WEEDS

All types of soil suffer if plants are cultivated on it in monoculture; that is to say, if there is always only one species of plant growing on it at one time. Healthy bacteria sicken, the soil becomes tired and poisoned, the animal and plant life in the soil deteriorates, until gradually no more viable crops can be harvested without the use of fertilisers and pesticides. The ensuing soil exhaustion is not merely the result of a shortage of minerals, but is also caused by exudations from the roots of the plants being cultivated. Oats, for example, make the soil acid.

When cultivating monocultures, the use of 'companion plants' (also known as weeds) can be very beneficial. This astonishing fact was known to a few biologists and agricultural experts in the past. Arable crops and weeds often enter into a kind of symbiosis to maintain the quality of the soil. For instance, charlock and wild radish, the companions of oats, de-acidify the soil and counteract the acidifying effect of the oats.

You should bear these observations in mind when combating weeds. Perhaps you can learn to use their beneficial qualities, such as gathering and drying nettles, in order to make use of their great curative powers.

Of course not every weed is a medicinal herb, and there are often good reasons for wishing that once a plant has been weeded or pulled up it will never return. Here is a helpful tip for unwanted weeds. The most suitable moment for pulling up weeds is when the moon is waning, ideally in the sign of Capricorn (Capricorn occurs in the waning moon from January to July). An especially favourable day is June 18th, up until noon (until 1 p.m. summertime). All shrubs and weeds that are removed during these few hours will not grow back and even the roots will rot away.

We have already acquainted future 'moon gardeners' with the old trick of taking advantage of the days when the moon is waxing in Leo, which are unfavourable for dealing with weeds. During Leo, all kinds of weeds shoot and sprout whenever they are 'touched'. Simply hoe over the newly laid out plot in Leo when the moon is waxing. Even the most delicate seeds will open and can then be weeded when the moon is on the wane in Capricorn, after which the plot will present a weed-free appearance for a long time to come.

In autumn you should weed all the plots for a final time when the moon is waning. This is a good preparation for the coming year.

You will hear of one or two other rules like this that are not dependent on the phase of the moon, particularly in connection with wood-cutting. They cannot easily be explained and can only be proven by trying them out for oneself.

Cutting Plants, Hedges and Trees

RULES FOR CUTTING BACK PLANTS

Pruning is one of the trickier tasks in the garden. We frequently learn by experience that the same expenditure of effort and expertise produces totally different results. One time the plant shoots up; then on another occasion it becomes stunted, spreads out over the ground, or even withers completely away.

The correct moment for cutting back a plant is when the moon is waning, or else when it is descending (from Gemini to Sagittarius). Plants and trees do not come to any harm if they are cut back when the moon is on the wane, because the sap does not escape. They cannot 'bleed to death', as the sap is descending.

PRUNING FRUIT TREES

Pruning fruit trees and shrubs is an important yearly task. However, many garden-lovers, including the 'pros', have from time to time had bad experiences with this. In some years it works, whereas in others it's a case of 'Murphy's law'. And no wonder, for this is a job that requires somewhat closer attention to correct timing.

The right moment for pruning fruit trees and shrubs is when the moon is on the wane, preferably on a fruit day in Leo, Sagittarius, or Aries. Equally suitable is the period of the descending moon, from Gemini to Sagittarius, because the tree sap is not ascending and will not exude from the cuts.

The most unfavourable time for pruning is when the moon is waxing, on a leaf day in Cancer, Scorpio, or Pisces. The tree loses too much sap and this inhibits the growth of fruit. Admittedly, the fruit tree will not be destroyed, but the crop yield will diminish or sometimes fail altogether. However, if the pruning is carried out precisely when the full moon is in Cancer, then there's no longer even any guarantee that the plant will survive.

GRAFTING

One of the more difficult jobs in the garden is the improvement of fruit trees through budding or grafting. Grafting, or binding a high-grade fruit-bearing or flower-bearing plant–shoot on to the stock of a common, but prolific base plant, is usually done to encourage healthy and vigorous growth, coupled with an increased power of resistance. It is a task that normally only very able gardeners dare to take on. However, anyone can manage it, provided they observe the following rule. Grafting on to fruit trees should take place when the moon is waxing, preferably close to full moon and on a fruit day in Aries, Leo or Sagittarius.

The sap of the tree quickly climbs into the new plant-shoot and links it more effectively with the root system below ground. It is best to perform this task on a fruit day and you will discover that the tree will bear fruit every year. If for reasons of time you are prevented from carrying out this job during the waxing or full moon, you should choose the ascending moon, preferably on a fruit day in Aries, instead.

AN ALMOST INFALLIBLE TREATMENT FOR SICK PLANTS AND TREES

On July 12th, 1984, a devastating hailstorm struck Munich and its environs, causing damage running into the billions. To this day there are dented cars on the streets of Munich bearing witness to the force of the egg-sized hailstones. There was also long-term damage that only came to light months later. Many conifers had lost their tips, which had been shot off by the hail. Subsequently, they slowly began to rot, starting at the top, until finally the trees died. In many cases the connection was not recognised and people ascribed the death of the trees to what is known as 'forest death' (as a consequence of general pollution). This is in fact a highly misleading expression, for the forest is not dying by itself. It should really be called 'forest killing'. And the solution to the puzzle? Precisely on July 12th, 1984, there was a full moon. So the weather had struck right at the most unfavourable moment, because the removal at full moon of the tip of a conifer tree or the tips of several branches of a deciduous tree, can seriously damage healthy trees or even cause them to die. Sick trees are condemned to death. They rot from the tip downwards.

The correct moment for lopping off the tips or trimming the branches of plants and tress that have stopped growing or are diseased is either on, or shortly before the new moon. Most plants can be treated successfully by removing their tips when the moon is on the wane, in the fourth quarter, or best of all at new moon. The tip should be removed just above a side-branch, which will then push upwards and adapt itself to becoming a new tip. In the case of flowers, shrubs and fruit trees, one sometimes has to cut away more than just the tip. This will encourge new shoots to form and the plant to blossom and bear fruit each year.

Where I come from, we use lengthened tree-pruners to cut off the tips of sickly or stunted trees, almost always with success. I am certain that the many sickly trees in our forests could be helped a lot if people would only take this simple measure.

In the case of flowers, shrubs and fruit trees, one sometimes has to cut away more than just the tip. With one of my own fruit trees ruthlessly I chopped everything away just a short distance above the grafting area. The tree put forth new shoots and since then it has blossomed and borne fruit every year.

As to the question of whether cutting off the tips at new moon could stop 'forest killing', I'm not sure about that, because this would not eliminate the causes. What is certain is that the results exceed all expectations. All sick trees that I have treated in this way have become healthy again.

Virgo Days – Workdays

Virgo	when the moon is **waxing**	when the moon is **waning**
	March to September	September to March

In the garden and the countryside the zodiac sign Virgo plays a very special role, as you may already have gathered from the advice given so far. Virgo is the ideal sign for planting and sowing, but there are also one or two other tasks for which it is favourable. Mountain foresters still know a great deal about the importance of these days. Tree cuttings can be planted without much difficulty during Virgo. At fairly large intervals a slit is dug with a spade; the cutting is put into it and the soil simply trodden down. If one or two saplings refuse to grow properly, then their little tips are nipped off at new moon. The trees will grow rapidly and robustly and it is often unnecessary to fence them in to protect them from damage caused by browsing wild animals. Where I live, as far as I know, saplings are not fenced in.

Plants such as geraniums that are re-potted during Virgo have the best prospects of becoming gorgeous, healthy balcony plants. Cuttings take root quickly in autumn, because then, Virgo is always in the waning moon. Cuttings can also be planted in spring, when the moon is waxing. This is particularly appropriate for geraniums that have been inside too long during winter. After their 'winter sleep' the geraniums are re-potted or divided during the Virgo days or else cuttings are simply planted from them.

Likewise, a lawn sown during Virgo, or even better during Leo at the time of the waxing moon, becomes a feast for the eyes. Town councils could save themselves large sums of money if they would pay attention to these times when laying out parks and lawns. The grass grows more vigorously and is much more resistant and a second sowing often turns out to be unnecessary.

One major exception to the Virgo rules is provided by lettuce. If it is planted in Virgo it shoots up without forming a head.

Virgo days are also suitable for another job in the garden – erecting or renewing fences. However, this should be done also when the waning moon or new moon are in force. Posts driven in at this time automatically stay firmly lodged. In any case, repair work of this sort usually comes up in the autumn or early spring, when Virgo occurs in the waning moon.

Of course the moon only remains for two or three days in Virgo each month, but whatever you manage to do by way of planting work during these days will be well worth the effort. In particular, anyone who has in mind a complete reorganisation of the garden would be well advised to carry out all planting and transplanting work during Virgo days when the moon is waxing. The only plants that do not need to be planted in Virgo are those that take root easily anyway and it is sufficient to plant these at the time of the waxing moon, or the descending moon. Set yourself priorities and during Virgo focus simply on all the plants that are problematic, and you will come through with flying colours.

Feeding Plants

In the pursuit of learning, every day something is acquired.
In the pursuit of Tao, every day something is dropped.
Less and less is done until non-action is achieved.
When nothing is done, nothing is left undone.
The world is ruled by letting things take their course.
It cannot be ruled by interference.

LAO TSU

GENERAL RULES FOR USING FERTILISER

Excessive use of fertilisers, which is nowadays the rule rather than the exception, prevents the normal formation of roots, especially in the case of fruit trees. The amount of fertiliser used should depend on the plant's requirements and is generally far less than is commonly assumed these days, particularly if one pays attention to the correct moment for putting on fertiliser.

As with everywhere in the garden and field, instinct and common-sense should be the yardstick, rather than rules, dogma and expert opinion. Good compost and dung, for instance, are still unbeaten as fertilisers, especially for fruit trees.

However, refraining altogether from spreading fertiliser is only helpful in rare cases. This method is only appropriate if you really know how to till the soil in an expert fashion. One farmer we know has not been using any fertiliser on his vegetable and cereal fields for the last ten years and has produced very good yields of high quality crops. When asked about his method, he said: "Taking it slowly is the whole secret." Apparently he had learnt his method from an old farmer. During every dry period in the vegetation phase he tills the soil in a variety of ways, using a piece of equipment that he developed himself. He only tills as long as the soil is warm and only goes down as deeply as the warmth reaches. When the moon is waxing, he only tills shallow furrows, while when it is on the wane, he goes deeper.

This method automatically leads in the course of time to the soil being 'touched' by all twelve energy impulses. Could it be that a basic agricultural principle for the next millennium lies hidden here?

An important observation of the natural world seems to have slipped completely into oblivion. During the period of the waning moon, starting after the day of full moon, the earth is able to absorb much more liquid than when the moon is waxing.

Not all that long ago, I was listening one morning to a discussion on the radio between environmentalists and representatives of the farming community. The argument raged this way and that, for and against the use of fertiliser and the importance of protecting the groundwater came up as an issue. There was no agreement in sight. I wanted to phone in and tell the participants that both sides were right, but then I decided against it.

It really is so simple. If the fertiliser is spread at certain times, it works itself into the soil, can be useful for the plants and does not go into the groundwater. At other times, the fertiliser stays lying on the surface, and because the materials are not taken up by the soil they pass straight into the groundwater and contaminate it.

Whenever possible, one should avoid putting down fertiliser when the moon is waxing, because the soil cannot absorb it and this merely pollutes the groundwater, making it unsafe to drink. Whenever possible, fertiliser should be put down at full moon or when the moon is waning.

Every farmer and gardener has discovered in the course of his daily work that on some days spreading fertiliser has devastating effects – the turf is scorched, roots atrophy or die and on other days the fertiliser produces the desired result, and there are no harmful side-effects. So, next time you put down fertiliser, pay attention to the position of the moon and observe how well the soil absorbs the substance in question when the moon is on the wane. This applies just as well to indoor and balcony plants.

Farmers and gardeners with complex work schedules may find it difficult to reconcile setting aside the day of the full moon as a day for putting down fertiliser. Although the period of the waning moon is quite long enough even for larger businesses to be able to take advantage of it. However, keeping to this rule shouldn't present a problem for allotment holders.

You will be surprised at the results you achieve by selecting the correct moment. By gradually weaning your plants from excessive dependence on fertilisers you will feel vindicated by your successes.

FERTILISING FLOWERS

When putting fertiliser down on flower beds, you should pay attention to the signs of the zodiac. This should not only be done at the waning moon, but on leaf days in Cancer, Scorpio or Pisces. For flowers with weak roots, you should select a root day in Taurus, Virgo or Capricorn. Flowers that do not seem to want to bloom any more should be given fertiliser occasionally on a flower day in Gemini, Aquarius, or Libra. Not too frequently, though; otherwise it's an open invitation to aphids.

CEREALS, VEGETABLES AND FRUIT

Cereals, vegetables and fruit should not merely bloom well but they should bear fruit that is full of vitality. The most suitable time for putting fertiliser down on this produce is during the fruit days of Aries and Sagittarius, when the moon is full or on the wane.

COMPOST HEAPS – RECYCLING THE NATURAL WAY

Since good quality, fully matured compost is one of the best things produced in the garden, not merely as a good fertiliser, we ought to discuss it in some-what greater depth. Whatever one receives as a gift from nature, one gives back to her, so that her inner strength and wholeness is maintained.

Making compost is one of the oldest forms of recycling. This chapter may not have much new information to offer skilled gardening buffs, but the con-stantly growing problems caused by rubbish have inspired many novices in recent years to try their hand at this art. Detailed descriptions are to be found in many gardening books. So we shall confine ourselves here to one or two tips.

The correct place for a compost heap is where it is protected as much as possible from the wind and lying in half shadow, in order to avoid it drying out. However, there should be enough warmth for rotting to take place. A position that is too shady slows down the process of transformation. When you have decided on a place, the surface of the ground where the compost heap is to go should be loosened up, to a depth of about four inches (ten centimetres). Another suitable alternative to this is to lay a four inch thick bottom layer of some dry, absorbent material, such as dry grass cuttings, twigs cut up small, mulch or straw. The base layer is then covered with loose, bulky material. The ground must not be covered with concrete or plastic or sealed in any other way. That would only lead to putrefaction and a build-up of moisture and would prevent worms coming up into the compost heap from below.

Starting up the compost and constructing a framework of planks around it should take place when the moon is waning. The material should be stamped tight when the moon is waxing, preferably one or two days before full moon. Alternatively, the compost heap could be started off when the moon is descending. If these times are adhered to, rotting takes place considerably faster. So at least one of these impulses should be borne in mind.

Now it is possible to start building up the compost. Place organic material and rubbish loosely, layer by layer, one on top of the other. All decomposable material from plant or animal rubbish that does not contain harmful substances is suitable for compost heaps. Branches should first be chopped up small. Diseased parts of plants do not belong in compost. Similarly, not all kitchen rubbish is suitable. The compost heap shouldn't become a rubbish dump. For instance, left-overs from cooked meals have no place on the compost heap and sooner or later they will attract undesirable vermin or even rats.

On earth days, especially in Virgo, but also in Taurus and Capricorn, attention should be paid when adding biological rotting agents such as powdered stone. Lime additives promote the formation of humus and healthy rotting. As an aid to the rotting process one can mix in half-ripe compost or garden soil between the individual layers and insert bulky material. When the moon is waxing, tramp it all down several times. Grass cuttings should never be piled too high (up to 2 to 4 inches/5–10 centimetres) because otherwise they will rot. Manure is suitable for providing additional enrichment with nutrients. Dry material can be moistened a little before being piled up. A rule of thumb for building up layers is to place dry material on top of damp and coarse material on top of fine material.

Good quality compost may well have a pleasant smell, but even so, you shouldn't put your heap right next to your neighbour's sitting area. Your compost heap can be sheltered effectively by means of a hedge or a row of runner beans.

If you observe these rules you will be able to harvest wonderful, ripe compost, which provides the best garden soil and fertiliser.

Harvesting, Storing and Preserving

Since time immemorial, methods for storing and preserving fruits from garden, field and forest have been employed. These techniques of fermentation, salting, smoking, boiling, roasting, drying and many more, allowed our forefathers to survive hard winters. More significantly, by paying attention to the correct timing in the lunar calendar when harvesting and conserving, our ancestors were able to achieve excellent results.

So often, measures to store and preserve food lead to variable results, although in each case the same rules of cleanliness are used. Almost every housewife has observed that now and then a jar of jam spoils after being opened for only a short time, whilst sometimes it can stand for weeks on the breakfast table and still taste like it did on the first day. Even without being opened, preserved fruit or home-made jams keep for varying lengths of time. Perhaps you will find the solution to this riddle when you have become acquainted with the rules for harvesting and storing.

The most favourable time for harvesting, preserving and storing, is when the moon is ascending, from Sagittarius to Gemini. The most suitable days for harvesting and storing cereals, vegetables and potatoes are in Aries. Fruit and vegetables are juicier when the moon is ascending, and they stay that way if they are harvested then and also have the best chance of tasting good and keeping for a long time.

The preservation of jams and juices should also be done when the moon is ascending. The fruit is much juicier and the aroma is much better, too. Jams keep longer and one can happily dispense with artificial setting agents or chemical additives (this also applies to preserving and bottling other foodstuffs).

Pisces days are an exception granted that they occur in the ascending moon. This period is not suitable for storing and preserving fruit and vegetables. Anything harvested at this time ought to be set aside for immediate consumption.

If you are not able to wait for a favourable time, you should take care at least to steer clear of the most negative influences.

Garden and field produce harvested when the moon is waxing should be consumed as soon as possible, if the moon is not currently in a sign which has an ascending force.

Virgo days should be avoided at all costs when harvesting, storing and preserving. Preserved foods, for instance, can very easily start to go mouldy. Cancer is not especially suitable, either, so anyone observing the ascending moon steers clear of this sign.

Arable crops and herbs that are to be dried should always be gathered and harvested when the moon is on the wane.

One should only clean cellar shelves when the moon is on the wane (during an air or a fire sign). This will keep them dry and prevent mould forming.

The Signs of the Zodiac in Garden and Field

ARIES   Aries days are fruit days with ascending force.

Very favourable: 🍃 sowing and planting anything that is supposed to grow fast and is intended for immediate use

🍃 grafting fruit trees (when the moon is waxing)

🍃 harvesting and storing cereals

Favourable: 🍃 planting and sowing fruit

🍃 cultivating cereals (when the moon is waxing)

🍃 fertilising cereals, vegetables and fruit (must be when the moon is waning or at full moon, April to September)

🍃 pruning fruit trees and bushes (when the moon is waning)

TAURUS   Taurus days are root days with ascending force.

Very favourable: 🌿 sowing and planting trees, bushes, hedges and root vegetables. Everything grows slowly and lasts well; harvest produce is especially suitable for storage.

Favourable: 🌿 setting up a manure or compost heap (when the moon is waning, May to October)

🌿 combating vermin found in the soil

🌿 occasionally putting down fertiliser for flowers with poorly formed roots

🌿 preserving and storing root vegetables (e.g. potatoes, carrots etc.)

GEMINI   Gemini days are flower days and the crossing point between the ascending and descending forces.

Very favourable: 🌿 planting and sowing any creeping or climbing plants

Favourable: 🌿 planting and sowing flowers

🌿 combating pests

🌿 occasionally putting down fertiliser for flowers that no longer bloom properly

CANCER Cancer days are leaf days with descending force.

Very favourable:
- ⬦ setting and sowing leaf vegetables (lettuce planted when the moon is on the wane forms a good head)
- ⬦ combating pests above ground

Favourable:
- ⬦ mowing lawns (even better when the moon is waxing)
- ⬦ watering indoor and balcony plants
- ⬦ putting fertiliser around flowers

Unfavourable:
- ⬦ setting and sowing plants that are to grow tall
- ⬦ pruning fruit trees and bushes (when the moon is waxing, especially in spring. Cancer at full moon is particularly unfavourable)
- ⬦ storing and preserving in the cellar is also unfavourable

LEO

Leo days are fruit days with descending force. Leo is the fieriest and most parching sign in the whole zodiac.

Very favourable:

- gathering herbs that strengthen the heart
- pruning fruit trees and bushes (when the moon is waning, suitable days for winter cutting)
- best day for cultivating cereals (when the moon is waxing) on wet fields

Favourable:

- sowing lawns (when the moon is waxing)
- planting fruit, but nothing that requires a large amount of water (tomatoes, potatoes)
- planting vegetables that are highly perishable
- planting trees and bushes
- grafting fruit trees (when the moon is waxing in spring)

Unfavourable:

- using artificial fertiliser

VIRGO Virgo days are root days with descending force. They are the best days for almost every type of work in garden, field and forest that is connected with setting, transplanting and new planting.

Very favourable:
- all planting and sowing work. The soil lets everything open up beautifully
- planting single trees that are meant to grow very tall
- planting hedges and bushes that are meant to grow very fast
- transplanting old trees (spring or autumn)
- re-potting and planting new balcony and indoor plants
- sowing lawns (when the moon is waxing)
- planting cuttings (when the moon is waxing, in autumn when the moon is on the wane)

Favourable:
- setting up a manure or compost heap (when the moon is waning)
- spreading all types of fertiliser
- combating vermin found in the soil
- occasionally putting down fertiliser for flowers with poorly formed roots
- erecting fence posts
- spreading manure

Unfavourable:
- planting lettuce (it runs to leaf)
- making pickles and preserves, storing

LIBRA

Libra days are flower days with descending force. Libra is a neutral sign, and there is hardly any task in the garden that is particularly affected either favourably or unfavourably.

Favourable:

- ⚲ sowing and planting flowers and flowering medicinal herbs
- ⚲ occasionally putting down fertiliser for flowers that no longer bloom properly

SCORPIO

Scorpio days are leaf days with descending force.

Very favourable:

- ⚲ sowing, planting and also harvesting and drying every kind of medicinal herb
- ⚲ combating slugs and snails (when the moon is waxing)

Favourable:

- ⚲ setting and sowing leaf vegetables
- ⚲ mowing lawns
- ⚲ watering indoor and balcony plants
- ⚲ spreading fertiliser for flowers and meadows (not so good for vegetables)

Unfavourable:

- ⚲ pruning fruit trees and bushes (when the moon is waxing, especially in spring)
- ⚲ felling trees (danger of bark beetle)

SAGITTARIUS   Sagittarius days are fruit days and the crossing point between ascending and descending forces.

Very favourable: ✎ planting and sowing all fruit and all vegetables that grow tall (runner beans, hops, etc.)

Favourable: ✎ pruning fruit trees and bushes (when the moon is waning in spring)

      ✎ cereal cultivation, particularly maize

      ✎ putting down fertiliser for cereals, vegetables and fruit in spring (must be when the moon is on the wane or at full moon)

      ✎ combating pests above ground

Unfavourable: ✎ hoeing and harrowing (weeds tend to shoot up)

      ✎ planting lettuce (also tends to shoot up)

CAPRICORN   Capricorn days are root days with ascending force.

Very favourable   ◢ harrowing weeds (when the moon is waning)

Favourable: ◢ planting and sowing root vegetables and winter
vegetables

◢ clearing and thinning out plants, forest edges and
hedges (when the moon is waning)

◢ setting up a manure or compost heap (when the
moon is waning)

◢ combating vermin found in the soil

◢ occasionally putting down fertiliser for flowers with
poorly formed roots

◢ preserving and storing root vegetables (for instance,
slicing up sauerkraut, when the moon is on the wane.
When the moon is waxing the fermentation process
takes place too quickly)

AQUARIUS

Aquarius days are flower days with ascending force. However, they are rather unsuitable for almost all gardening tasks. One should confine oneself to essential work. In garden, field and forest Aquarius is a somewhat infertile sign.

Favourable:

- hoeing and harrowing; the weeds can be left to rot
- occasionally putting down fertiliser for flowers that no longer bloom properly

Unfavourable:

- planting seeds, because the transplanted seedlings will not take root and so they will die

PISCES

Pisces days are leaf days with ascending force. Everything harvested on these days ought to be consumed at once.

Favourable:

- planting and sowing leaf vegetables
- watering indoor and balcony plants
- mowing lawns
- fertilising flowers
- planting potatoes when the moon is waning (especially good when Pisces falls on the third day after full moon)

Unfavourable:

- pruning fruit trees and bushes (when the moon is waxing, especially in spring)
- preserving and storing

Correct Timing in Farming and Forestry

Three things are needed in order to defeat any adversary:
to be glad when he is right,
to be sad when he is wrong,
and never to behave foolishly towards him.

Four things are needed to save the world from humanity:
accept the ignorance of others
and spare them your own.
Give to them from your substance
and expect no part of theirs.

INDIAN PROVERB

Farming has undergone enormous changes in recent decades, both voluntary and enforced, for better and for worse. Large-scale farming relying on rigid schedules cannot easily accommodate a return to an observance of lunar rhythms. But perhaps after reading the previous chapter you will have gained a deeper insight into the dynamic cycles of nature, which have not changed at all for thousands of years. You may have even found a plausible explanation for many a puzzling experience you encountered when sowing, planting, harvesting and storing.

For many small farmers a knowledge of the natural cycles is of great interest and also has practical applications. An increasing number of the public are now prepared to go a long way in search of healthy cereals and meat and 'milk from happy cows'. As the demand for healthy food (organic produce free of pesticides and chemical fertilisers) increases those farmers with smallholdings are in the strongest position to convert to this type of farming.

All gardeners, farmers and foresters are in the same boat. They profit from the harmonious interplay of heaven and earth, of sun, wind, clouds, water and warmth. It is not entirely their fault that some of the laws that prevail between heaven and earth have apparently slipped into oblivion and that this has led to many problems – the excessive use of fertilisers, the poisoning of the soil, the contamination of the groundwater, and a less healthy harvest.

However, the signs that things are taking a turn for the better are on the increase. It has become clear to many people that the price of disregarding natural cycles is much higher in the long-term than the short-term gains to be had in terms of agricultural and livestock yields. The North American Indians knew this from the outset:

> *Only when the last tree has been cleared,*
> *the last river poisoned,*
> *the last fish caught,*
> *will you discover that you cannot eat money.*
>
> NATIVE AMERICAN PROVERB

And it is especially in their country, which has been so severely exploited by maize-farming and other monocultures, that the march has begun towards the agriculture of the future. Large concerns have been transformed into many little ones; in the truest sense of the word environmental protection is gaining ground, fields are again being surrounded with hedges and avenues of trees.

Perhaps it is helpful to remind ourselves once more that it was precisely the ancestors of present-day farmers and foresters, especially woodland and mountain farmers, who discovered, preserved and passed on the knowledge of natural rhythms.

The rules presented in this book certainly cannot be applied overnight to agriculture and forestry. Making changes is a slow process, but one which will only get under way when the will and intention are there. Here is a suggestion for staring on the right track.

Simply reserve one or several small plots, and try out all the rules of plant cultivation that are introduced in this book. Do not make any changes in your other work routines. Just observe how these little bits of land develop.

Lunar Cycles in Farming

CEREAL CULTIVATION

If someone sowing cereals manages to catch the influences of both the waxing and descending moon, and on top of that a fruit day in Leo or Sagittarius, then the only thing left that could wreck his plans would be the weather.

If it happens to be raining cats and dogs on these days that are ideal for planting, no farmer will venture into the fields. Thank heaven that nature is arranged in such a way that not everything always fits neatly together.

Aries and Sagittarius days are especially suitable for cultivating cereals, and Leo days are a second best. However, if the fields are very dry, there is a danger that during Leo the soil may become even more parched. Leo days are best when the fields are already very moist.

The best time for storing cereals is on a fruit day in Aries.

LOTTERY DAYS

Perhaps you may have wondered why a particular variety of cereal turns out especially well in some years and particularly badly in others. Of course wind and weather, seed quality and soil characteristics play a part. Yet sometimes one farmer will succeed in bringing in a good harvest, while his neighbour, working under the same conditions, only achieves a miserable yield.

From the knowledge you now have of lunar rhythms and their influence on sowing, fertilising and harvesting, you will probably realise that one main reason for the differing yields is to be found here.

In this connection you should be made aware of a curious rule which almost every farmer knew in previous times, and which according to our own personal experiments is still entirely valid. It concerns two days in the year, namely July 8th and July 20th. Anyone wishing to find out which type of cereal will produce the best harvest the following year should sow a few grains of each cereal variety in the ground. The seeds that have sprouted the best on July 20th will flourish the best in the coming year.

Try this rule out in a test bed. The results will speak for themselves. These days belong to the so-called lottery days. The origin of their name lies in the fact that these days were once seen as a sort of lottery. Depending upon what happened on these days, people predicted future events, principally connected with the way the weather would develop. There are a number of farmers who still observe these days, especially the weather rules connected with them. However, climatic changes that have occurred over the past few decades have jeopardised their effectiveness.

PROCESSING MILK

Almost every form of milk processing is at its most successful on fat days in Gemini, Libra and Aquarius. The warmth days of Aries, Leo and Sagittarius are also good moments to carry out this work. Neutral days, when neither beneficial nor harmful influences will prevail, fall in Taurus, Virgo and Capricorn. Unsuitable days occur in Cancer, Scorpio and Pisces. On these days, one can churn and churn and no butter will form.

When making cheese, the phase of the moon plays an important part, but this all depends on what sort of cheese is to be produced. If the cheese is to mature quickly, then the waning moon is more suitable while for slow-maturing cheeses the waxing moon is more favourable. In general, Virgo is not a favourable sign for cheese production. Fruit days are more suitable for this.

PUTTING DOWN FERTILISER

As we have already mentioned, the earth is able to absorb more liquid when the moon is on the wane than when it is waxing. Recently there has been much discussion in the media because ministries of agriculture all over the world are having to budget colossal sums for the protection of water resources, as a result of the pollution of groundwater and rivers with excess fertiliser – phosphates and nitrates from agriculture and waste water. A large part of this money could be saved if attention were paid to lunar cycles when using fertilisers.

The fact that the absorbency of the earth varies is evident in other ways. Have you ever noticed that flooding is much more common when the moon is waxing? At this time the earth is unable to absorb so much water. On the other hand, steep mountain slopes are much more vulnerable to landslides when the moon is waning, because the earth is wet and heavy and saturated – particularly if there are no healthy trees firmly anchored in the soil, or if the mountain forest is diseased. This fact is of great importance for modern agriculture and forestry and for the protection of water resources and should be looked at urgently.

Wherever possible, fertiliser or manure should be spread when the moon is waning. With cereals, vegetables and fruit, ideally on fruit days in Aries, Leo or Sagittarius, otherwise during Virgo or on another earth day in Taurus or Capricorn. Liquid manure should be spread if possible when the moon is full, or alternatively, when the moon is on the wane. Then the groundwater will remain unharmed. Never spread artificial manure during Leo. The plants will be scorched because Leo has a severely desiccating effect.

One might argue that if every farmer were to spread liquid manure at the same time, the stench would cause a great nuisance. And yet it is surely preferable to put up with the stink for three days rather than poisoning the groundwater.

When using fertiliser be guided by instinct and common-sense. Fertiliser manufacturers are not free from self-interest and often exaggerate the quantities required. Good compost and manure are still unsurpassed as fertilisers, especially for fruit trees.

HARVESTING AND STORING CEREALS

Good results can be obtained when harvesting and storing cereals if the farmer is able to do this during the ascending moon, or, alternatively, when the moon is on the wane, especially during Aries or on another fruit day. The cereal will last much longer and be less susceptible to beetles and mould. By using these natural rules, vast quantities could be saved on pesticides.

All water days are unsuitable for storing and should be avoided.

PATHS AND WATERCOURSE MANAGEMENT

Many architects and builders have learnt through experience that paving stones laid out of doors sometimes become wobbly after a short time and that verandas or paths covered with gravel become uneven, despite having being laid with care and expertise. On another occasion, the path stays as firm as if it were set in concrete. Here, too, the timing of the work is of decisive importance for a successful outcome.

Regarding the rules of watercourse management, observe the variable effect of flood water in the bed of a stream. When the stream is in spate during the waxing moon, the water leaves behind a lot of gravel in the bed of the stream and the stream overflows its banks more often. When the moon is waning, the stream washes and carries the gravel away.

Country tracks should be laid or filled with gravel when the moon is on the wane. Capricorn days are especially suitable for this. If the work is done when the moon is waxing, the surface remains loose, subsides or becomes rutted. The very first rain will wash the new gravel away.

Paving stones in the garden or on paths should be laid when the moon is waning. If they are laid out when the moon is waxing, they eventually become wobbly. This is important especially with the paving around entrance gates, because the immense weight of vehicles rapidly dislodges the paving stones or often even breaks them.

Excavating springs or searching for water should be undertaken in the star sign of Pisces, when the moon is waxing, if possible in the second quarter, close to full moon. Never work on a spring when the moon is on the wane or descending. The water will disappear and find another path for itself.

Building up the banks of streams and rivers, by embedding rocks and timber, should be done during a water sign in Cancer, Scorpio or Pisces, when the moon is waxing. If this work is carried out when the moon is waning, the built-up bank will be undermined and washed away and the river will overflow.

Work on drainage and sewage and repairing water pipes is similarly most successful when the moon is waxing during a water sign.

FENCING

The most suitable time for erecting or renewing fences is when the moon is on the wane, ideally on the day of the new moon. Posts driven into the ground at this time automatically remain firm and nails placed in wood stay firmly in place, particularly during the earth-days of Virgo, which in autumn are always in the waning moon (from September to February). Alternatively the descending moon is suitable. However, the closer to full moon, the less favourable it is.

SETTING UP HAYSTACKS

The best time to set up a haystack is when the moon is on the wane. The hay remains well ventilated and dry, does not rot and the danger of spontaneous combustion is much less. If the moon is waxing the hay goes grey and mouldy.

CARE OF ANIMAL DWELLINGS

Every farmer and farmer's wife knows that careful maintenance of stables, cowsheds, and pigsties can be a tedious and time-consuming task. However, those who observe the following advice will be able to save themselves a great deal of drudgery.

Watch out for opportune moments at the waning moon, or alternatively the ascending moon and for an air sign in Gemini, Libra or Aquarius. The animals' quarters will remain cleaner longer, be protected from vermin and mould and dry quicker. Mould on the walls can be washed off more easily when the moon is waning (ideally using water with a little vinegar and a scrubbing brush). Persistent dirt can be removed particularly easily during a water sign in Cancer, Scorpio or Pisces.

Bedding straw for animals, too, should be prepared during the waning moon. Alternatively, if time is short, one should favour the ascending moon. Incidentally, a good underlay for bedding is straw or dry beech foliage.

Walls should be painted when the moon is waning. This task is particularly successful if carried out on an air day in Gemini, Libra or Aquarius.

DRIVING CATTLE OUT TO PASTURE

One of the decisive factors influencing the behaviour of animals in pasture is the day that they are driven out for the first time in the year. If one chooses the descending moon (especially during Libra) then the animals are happy to stay out in the meadows and do not try to return home before they have eaten their fill. You should also make sure that you choose a Monday, Wednesday, Friday or Saturday.

Cattle should never been driven out to pasture for the first time on a Tuesday or a Thursday. This rule is still observed in the Alps, particularly at the time of the annual journey up to and down from the alpine pastures. These two days are also unsuitable for moving cattle to other places or bringing them home from market. At these times, they are not so generous with their milk and subsequently will often not conceive. Leo and Cancer days should both be avoided. Driving cattle out on these days makes them wild and difficult to control. During Cancer they are forever returning to the cowshed door, especially cattle grazing on alpine pastures.

In autumn cattle should be put out to pasture for the last time when the moon is waxing.

CALVING

When cows are in calf round about new year, the calves are born around October. This is the best moment for calving and produces the healthiest animals. Rarely will the vet be called on to attend a birth at this time of year.

HEALTHY CHICKENS – HEALTHY HENS

For hens, too, there are equivalent rules. The best chickens are produced if the hens' eggs are incubated, whether by the hens themselves or in an incubator, so that the chickens hatch at full moon. The incubation period is always the same, so even in large-scale businesses it is perfectly possible to calculate the correct moment and stick to this rhythm.

MOVING ANIMALS INTO NEW QUARTERS

The period of the ascending moon is the most suitable time for moving animals into a new shed. The animals are less restless and happier to remain inside. The day of the week is also influential. Newly bought or sold animals should never be moved from one shed to another on a Tuesday, Thursday or Sunday. Even moving a stall-box is unfavourable on these days.

WEANING CALVES

Unfortunately, allowing calves to be suckled by the cow is no longer fashionable. Perhaps one reason for this is that the art of timing has been forgotten.

Start weaning calves shortly before full moon, and let them drink milk for the last time at the full moon itself. However, the full moon in Leo, Cancer or Virgo should be avoided. In Leo you will have yelling animals; in Cancer they will always be coming back for more; in Virgo they often get aggressive afterwards and their restless behaviour and brawls with other cows make them a danger for the whole herd, especially in the mountains.

Rules of Forestry

Wood is a wonderful substance. For thousands of years it has given humanity warmth and tools, protection and beauty, through the work of the great carvers. To protect and preserve the forests of the earth, the source of all this wealth is one of the most important tasks of our time. Fortunately, despite the death of our forests through pollution and clearance of the rain forest, there are still grounds for optimism. Positive measures are being taken by countries such as Sweden, where not a single tree is cut down without a new one being planted. World-wide afforestation and the efforts of environmentalists, who are gradually making their views felt, are equally beneficial. An immense contribution to saving our forests could be made by the rediscovery and application of the ancient rules of forest maintenance and tree-felling, in a variety of ways.

Wood is a substance that is full of life. Even after it has been cut down the wood lives on. It goes on 'working', to use the language of timber experts. Depending on the type of wood, the season and – as we shall see – the moment when the tree was felled, the wood will dry rapidly or slowly, stay soft or become hard, stay heavy or become light, develop cracks or remain unaltered, warp or stay flat, fall prey to rot and woodworm or remain protected from pests and decay.

Left-spiralling tree.   Right-spiralling tree.

As with all the other rules, there are no good and bad days for felling timber. The decisive factor is the purpose to which the wood in question will be put. It makes a big difference whether wood is intended for floors, barrels, bridges, roof trusses, musical instruments or carving. Of course, the type of wood also has to be taken into consideration, as well as its age and pattern of growth.

Trees grow either straight up, spiralling to the right, or spiralling to the left, as can be seen from the bark. The difference is not difficult to recognise. A rightwards spiralling tree grows upwards like a corkscrew held upright. This 'direction of rotation' must also be taken into account when using a particular piece of wood.

Roof shingles, for example, should either be straight-grained or spiralling slightly to the left. In wet weather, wood stretches; in sunshine, on the other hand, it twists very slightly, thus allowing dry air under the shingle.

With wooden gutters, which are sometimes still used, the opposite is true. The grain of the wood should run straight, or spiral slightly to the right, because right-spiralling wood stays put after it has been felled. That is to say once the wood is in place the rotation does not continue. Left-spiralling wood will cause the gutter to twist little by little and the water will run out.

The strange thing is that left-spiralling wood 'works', continuing to move, more than right-spiralling or straight-grained wood. Furthermore, lightning only strikes trees that spiral to the left – a useful piece of information if you are caught in a thunderstorm in the forest. You should only stand under a straight or rightwards-spiralling tree.

Even today in Tyrol and many other countries, the correct felling times are still observed. You will get a knowing wink from many timber merchants and be led to a particular stack of timber, if you insist on wood that has been cut at the correct time. Timber firms in many parts of the world, such as Brazil and the South Seas, conclude contracts in which the wood-cutters must only sell timber that has been felled at particular times.

Of course, many firms no longer pay any attention to the favourable moment, either for organisational reasons, because insufficient importance is accorded to the work, or else quite simply because the knowledge is no longer available. Moreover, paying attention to correct timing may appear at first sight to be complicated and costly, but this is not necessarily so. The work has to be done in any case. Anyone who looks out for wood that has been cut at the right time will find his labour rewarded many times over.

In the course of recent years I have got to know several people who had previously never heard anything about the rules regarding the correct timing for cutting wood, but who were prepared to put this ancient knowledge to the test. They were all surprised at the result and how unfailingly these influences still hold good today.

Paying attention to these rules could have excellent consequences for industry. Products such as furniture, bridges, buildings, tools, timber and many other things, would become much more durable, and wood preservatives would no longer be needed.

Our society that has grown so accustomed to discarding products may find it hard to appreciate the value of a sustainable culture. Those people at one with nature and who are in tune with their environment have long understood the value of natural laws. Of course everyone is happy to enjoy the advantages of science, technology and progress without thinking too much about the long-term consequences. But if the possibility exists of solving the problem of waste, if the destruction of forests can be slowed down, if we can release fewer poisons into our surroundings, then everyone ought to know about such possibilities and be able to take advantage of them.

Perhaps in the future architects, joiners, carpenters and interior decorators will unite together and commit themselves to using wood cut according to the old rules. They will recognise that environmental compatibility, quality and durability are becoming the most important factors for more and more people when deciding to buy.

In times like today, when ecological house-building is gradually coming to the fore, sufficient customers could be found who would know how to appreciate such things. Anyone who commissions the construction of such a house will be at pains to build in the most environmentally friendly manner. If his roof truss warps into curves after a few years or splits the wood, then even with the best will in the world he can be driven to despair. One can often observe how natural wood and materials, left untreated for the best of reasons after some years nonetheless have to be treated, at considerable expense. The good intention was there, but watching wood getting wetter and wetter and threatening to rot makes many a builder throw up his hands. All problems of this sort could be avoided if one used wood that had been cut in accordance with the position of the moon. (Prospective builders are urged to refer once more to the short section on paths and watercourse management. All the advice given there applies to laying paths, drives and paving stones, whether building a new house or simply renovating it, whether in the country or in the town.)

To anyone asking the perfectly legitimate question as to how one is to find wood nowadays that has been cut at the correct time, we can only suggest referring to the yellow pages. Ring up some timber merchants and ask if you can be told the time when the wood was cut or ask an association of timber merchants to send you their list of members.

Forestry regulations from previous centuries give precise directions concerning the times to cut wood, taking into consideration the lunar cycles. There is many a retired forester who can probably remember the tales that his own grandfather told about how meticulously forestry workers observed the most favourable times for cutting the various types of wood and drew up highly refined yearly plans for cutting timber. A yearly plan is necessary because the best times vary slightly from one year to the next and there are even some dates that only occur every two years or so. Timber preservatives were unknown, because the choice of the correct moment achieved the desired effects for the particular quality of wood.

THE CORRECT TIME FOR CUTTING WOOD

Almost everyone that has anything to do with cutting and processing timber knows that in general winter is the best time for obtaining wood. The sap has descended, and the timber will warp less after it has been felled. However, in addition to that, there are a large number of special dates which have a clearly discernible bearing on the characteristics of wood.

The following quite comprehensive set of rules comes down to us from very ancient times. The text under consideration is dated 1912. All the rules stated in this historic document are as valid as ever.

INDICATIONS FOR WOOD-CUTTING AND WASTING DAYS

By Ludwig Weinhold

Recorded by Michael Ober, master cartwright in St Johann, Tyrol,

copied by Josef Schmutzer, December 25, 1912

1. Wasting days are April 3, July 30, and St. Achatius day. Even better when these fall during the waning moon or on a Lady Saint Day.

    These days are also good for casting bullets and shooting.

2. For timber to remain firm and tight-fitting it is good to cut it in the first eight days after new moon in December, on a day with a weak sign.

    For straight-wood or making-wood (beech etc) to remain tight fitting and firm, it should be cut in the new moon in Scorpio.

3. So that the wood does not rot, it should be cut in the last two days of March, with the waning moon in Pisces.

4. So that the wood does not burn, there is only one day when it should be cut: March 1st, preferably after sunset.

5. So that the wood does not shrink it should be cut on the third day of autumn. At the start of autumn on September 24, when the moon is three days old, and on a Lady Day in Cancer.

6. Working on firewood so that it grows back well should be done in October during the first quarter of the waxing moon.

7. Timber for sawing should be cut down when the moon is waxing in Pisces. Then boards and timber will not be worm-ridden.

8. Timber for bridges and arches should be cut down when the moon is waning in Pisces or Cancer.

9. So that the wood becomes light, it should be cut down with the moon in Scorpio and in August. If it is cut down in Taurus, i.e. when the August moon has been waning for one day, then it will remain heavy.

10. So that the wood does not develop cracks or open up, it should be cut before new moon in November.

11. So that the wood does not split, it should be cut on June 24 between 11 and 12 o'clock.

12. Straight-wood or making-wood should be cut down on February 26 when the moon is on the wane. Better still on a Cancer day.

These indications have all been proven and tested.

Of course this set of rules needs interpreting if it is to be understood by every-one today. There now follows an explanation of these rules, as well as many additional tips, arranged according to the quality or type of wood desired or the intention being pursued on a specific date.

WASTING DAYS – GRUBBING AND CLEARING

Every commercial forest needs maintenance. Anyone who wishes to clear and clean up a wood or forest border, or who wishes to deforest and replant, should look out for wasting days, also known as grubbing days, on April 3rd, July 30th, and St. Achatius's day, on June 22nd. The results of the work will be even better if these days fall during the waning moon or on a Lady day. Lady days are holidays devoted to the Virgin Mary, such as the feasts of the Assumption or Purification. These days can be found in any farmer's calendar and include August 15th and September 8th. Trees and bushes cut down at this time will not grow back. Other days for grubbing and clearing are the last three days in February, if they fall in the waning moon. Wood cut at this time will not grow back and even the roots will rot.

WOOD FOR TOOLS AND FURNITURE

Wood that is cut in the first eight days after the December new moon in Aquarius or Pisces stays firm, does not warp or get dry and fall apart and main-tains its volume – an important feature for instance of skirting boards and edging strips. Wood that is used to make tools and implements and furniture should be cut at the new moon in Scorpio (usually in November). The bark should be removed immediately, as wood that has been cut in Scorpio, or uprooted before a storm is vulnerable to attack from beetles. Alternatively, it should be cut down on February 26, provided this date falls during the waning moon (which is not always the case) and especially if the moon is in the sign of Cancer.

NON-ROTTING, HARD WOOD

Non-rotting wood must be cut during the last two days in March, with the moon on the wane in Pisces. These days do not come round every year. So formerly people used to look out specially for them or else cut wood on alternative days such as New Year's Day, January 7th, January 25th, and from January 31st to February 2nd. Timber that is cut down on these six days will not rot or get woodworm. Furthermore, wood that is cut at New Year or between January 31st and February 2nd will become very hard as it ages. Alternative days are warm summer days when the moon is waxing and the wood's sap content is at its highest. This type of wood is suitable for building pile foundations in water, mooring piers for ships, and bathing platforms.

NON-INFLAMMABLE WOOD

Anyone who has visited a 'museum village' (such as Kramsach in Tyrol), with its centuries-old buildings, barns, equipment and tools, will have also certainly seen benches next to stoves, wooden pot-holders, bread paddles and wooden fireplaces. What is strange is that hardly anyone asks why the semi-circular pot-holders used for lifting red-hot pots and pans from the stove lasted for so long – for centuries even – without burning. Or why wood exposed directly to fire did not burn, such as wooden fireplaces and the wooden tools used to scrape the sides of lime kilns. Admittedly, the wood was blackened, but it neither burned nor glowed. Perhaps you, too, have come across a box of matches that absolutely refuse to burn. The solution to the puzzle is this. There are specific times when natural impulses ensure that wood is non-inflammable.

For instance, wood that is cut on March 1st, especially after sunset, is fire-resistant, regardless of the position of the moon and the sign through which it is currently passing. This is a mysterious and yet valid rule and whoever tries it out will find it confirmed. Many implements, farmyard buildings, barns, log cabins and mountain huts were built out of such timber in order to make them fireproof.

My parents' house, which was completely gutted by fire in 1980, was made of non-inflammable wood. The metal of some agricultural implements inside was subsequently found to be partially melted, so great was the heat. The building itself remained standing. The wood was only charred on the outside. When it was to due be broken up using a tractor, the timber would not give way. In the end the house had to be sawn up beam by beam. It then transpired that only a few millimetres of the outermost layer of the wood was charred. The interior had remained completely intact. Part of the timber was later reused in the construction of two new buildings.

NON-SHRINKING WOOD

Wood that does not shrink is needed for all kinds of purposes. Such wood is best cut on St Thomas's Day on December 21st, between 11 and 12 o'clock. This day is the best wood-cutting day of all. After this date, timber should, with certain exceptions, only be cut down in winter when the moon is waning. Other periods suitable for cutting non-shrinking wood are February evenings after sunset, when the moon is on the wane; September 27th; every month during the three days after new moon; Lady Saints' days (such as August 15th and September 8th) when these fall in Cancer. Wood that is cut at the new moon in the sign of Libra will not shrink and can be used straight away. Moreover, timber that is felled in February after sunset, will become rock-hard as it ages.

FIREWOOD

The best time for cutting firewood is the first quarter of the waxing moon in October. That is to say the first seven days after the October new moon.

In general, however, firewood should be cut after the winter solstice when the moon is on the wane. But the tree top should not be removed straight away, and in hilly conditions you should leave the timber pointing downhill for some time, so that the last of the sap can be drawn to the top .

TIMBER FOR PLANKS, SAWING AND BUILDING

The most suitable time for cutting timber for planks and sawing is the waxing phase of the moon in Pisces, because the wood will not then be attacked by pests. The star sign Pisces only appears in the waxing moon from September to March.

TIMBER FOR BRIDGES AND BOATS

Have you ever crossed a wooden bridge on a rainy day? It's as well to hold tight to the railings, as the wooden slats are often very slippery. On the other hand, there are old wooden bridges over mountain streams in the Alps that are safe underfoot, do not rot and seem to have been built to last forever.

The fact that nowadays Alpine clubs and tourist organisations obviously no longer pay any heed to such influences when building bridges, is something that every mountain hiker has been forced to discover. There would be no need for so many tourists to be fetched back with sprained limbs by the mountain rescue service, if more attention was paid to correct timing when cutting wood.

Timber for bridges, boats and rafts should be cut when the moon is on the wane, in a water sign in either Pisces or Cancer. The wood will not rot and will be safe under foot. This rule used to be observed in the choice of wood for wash-stands, which have to endure constant soaking without becoming slippery.

Timber for Floors and Tools

Broom handles and other wood for tools should be supple and firm to hold, not easily breakable, flexible and above all light. The best time for cutting this type of wood is during Scorpio days in August, which almost always come before the full moon.

If the wood is to be supple but solid, for instance, for wooden floors that will take a lot of strain, then it should be cut on the first day after the full moon in the sign of Taurus, which does not happen every year.

Non-splitting Wood

Timber that must not split, such as for furniture and carving, is best felled in the days prior to the new moon in November. Equally acceptable alternatives are March 25th, June 29th and December 31st. The top of the tree should be left to point down the valley or if the tree is on level ground it should be left on the tree a little longer, so that residual sap can be drawn to the top.

Wood that is to be used quickly, for instance when rebuilding after a fire, must under no circumstances split later. The best felling time is June 24th, between 11 and 12 midday (12 and 1pm summer time), which was once a special day. Timber workers would turn out in swarms and for a whole hour would saw for all they were worth.

The best bridge timber is obtained when the new moon coincides with Cancer. This rule is as valid today as it ever was.

Christmas Trees

Finally a tip for the season of peace. If fir trees are cut down three days before the eleventh full moon of the year (generally in November, but sometimes in December), they retain their needles for a very long time. These trees used to receive a 'moon stamp' from the forester and were somewhat more expensive than other Christmas trees. Spruce trees cut down at the same time don't shed their needles either, but should be stored in a cool place until Christmas. However, they still lose their needles earlier than firs.

A relative of mine has owned a Christmas tree like that for more than thirty years and it still has its needles. I myself still have the first advent wreath I made after moving to Munich in 1969. If I pluck off a few needles, they smell sweet even today.

It would be a good idea to gather shoots for advent garlands three days before the eleventh full moon, because in that way beautifully laid out advent tables won't get covered with needles. Of course, knowledge of this rule should not be seen as entitling you to go marching into the forest to 'poach' your Christmas tree there.

Obviously, one cannot always get a Christmas tree that was cut exactly three days before the eleventh full moon. And so here's another tip. Christmas trees and garlands also last longer and do not lose their needles so quickly if one watches out for the waxing moon.

Garlands of dried flowers also last longer if they are picked when the moon is waxing.

The Moon as a Helper
in the Home and
Everyday Life

Soft conquers hard,

weak conquers strong.

The flexible is always superior to the immovable.

This is the principle of controlling things

by bringing oneself into tune with them,

the principle of mastery through harmony.

LAO TSU

The fact that the lunar cycles have slipped into oblivion in the household and in city life is scarcely surprising. The wind that has been blowing for decades from the 'land of limitless possibilities' bears a very special message – freedom and self-realisation are basic rights that take priority over our obligations towards ourselves, our neighbours and nature.

The industrialised society of the Western world has divorced itself from nature. We live in a consumer society, using up natural resources without any thought for the long-term consequences upon our environment. Our priorities in all areas of life are speed, convenience and efficiency, to be achieved at any cost. Western civilised man has gradually become convinced that electricity comes out of the socket, that a caustic cleansing fluid vanishes into thin air when it reaches the waste-pipe and that the whole of life comes to him out of the television tube.

Doing the housework in harmony with the rhythms of the moon is easier and more enjoyable and reduces the already high level of stress that many people toil under. There are many tips for the daily household round, lurking in the preceding chapters (healthy diet, cooking, herbal knowledge, home improvement etc.), but there are many things still to be said and some things that are worth repeating.

Almost all housework which usually involves cleaning, removing and 'flushing out' is dealt with much more successfully and effortlessly when the moon is on the wane. Of course it isn't possible to postpone all work in the house to the waning phases of the moon, and simply to twiddle one's thumbs while the moon is waxing. However, if you can manage to move gradually a part of the workload to this period, you will be astonished at the effects arising from even minor postponements.

Let your own experience speak for itself. There are times when any kind of housework comes easily while at other times it simply never ends. In practice, one often notices, without being able to give a reason for it, that things such as furniture, floors, laundry, and windows become clean more quickly and easily than on other days. Sometimes everything goes really smoothly, while at other times everything goes wrong.

When you have actually tried out this rhythm you will be able to bring a lot of pleasure and energy to this time-consuming work. On some days many things seem to get done with almost no effort at all.

Whether a task goes smoothly or badly, whether the room or the washing is cleaned satisfactorily, depends on the moment when the work is done. The rule of thumb with household chores is when the moon is waning everything is easier.

Washing Polishing and Cleaning

WASHING DAY

Especially in large families, it is impossible to do all the washing that accumulates only when the moon is on the wane. Trusting in your powers of invention, you at least should try the experiment of putting off the bulk of the work until the waning moon. The results of this new approach will speak for themselves and awaken the inspiration to undertake further alterations in your working routine.

Paying attention to lunar rhythms when doing the laundry can produce particularly good results. However, you must be prepared to give up using powerful washing agents and excessive quantities of detergent. Otherwise you may not succeed in reproducing the observations that follow. The effect of the lunar cycles is so subtle that it may not seem plausible to claim that the moon's rhythms can produce such effects. Yet these forces do exist and can be of great use in the long run. When you try doing the laundry in harmony with lunar cycles, you will be discover a new energy and pleasure in this usually time-consuming work. On some days many tasks appear to be done with almost no effort at all.

For example, problem stains, can be removed much more quickly when the moon is waning than when it is waxing. When washing, water days in Pisces, Cancer and Scorpio produce far better results. Moreover, the environment will be spared, since waste water is broken down more easily when the moon is waning.

When the moon is on the wane, which is when I do most of my washing, I use only a quarter of the prescribed quantity of detergent. I don't have any problems with my washing machine getting furred up either. If I find any calcium deposit in the filter I simply add a little vinegar to the water. I really think that in this way I can make my personal daily contribution to environmental protection.

Try out a simple test. When the moon is waning, place a very dirty piece of clothing in a full wash basin and add some detergent or soft soap. Then, when the moon is waxing, do the same thing under the same conditions, and compare what happens. The results will astonish you. When the moon is waxing, lather will accumulate and stubborn stains cannot be removed. Whereas when the moon is on the wane, the dirt will come out effortlessly. Just by looking at the suds one can see where it has all gone. Perhaps you may have noticed that sometimes the laundry smells especially fresh and light, even though the washing procedure and the quantity of detergent used were the same. This is due to the beneficial influences of the waning moon. Doing most of the washing when the moon is waning also saves detergent and is less hard wearing on one's clothes.

Here is a tip for eradicating grease, especially car grease and stains from bicycle oil. When the moon is waning, on a water day, rub in a little lard. Then wash as usual. Environmentally conscious people often find that some stains cannot be dealt with by using natural methods. Such gallant fighters for a better environment will be glad to discover that paying attention to the phases of the moon can produce good results.

DRY CLEANING

Materials made of delicate fabrics or that stain easily, such as lambskin, leather, and silk, should only be taken to the dry cleaners when the moon is waning. The fabric will come to no harm at that time, the garment will last longer and the colours will not fade. If possible avoid the sign of Capricorn for dry cleaning as it causes the dreaded sheen to appear on garments.

Seasonal clothing should only be washed or cleaned when the moon is on the wane, before it disappears into the wardrobe for six months or more.

WOODEN AND PARQUET FLOORS

Wooden floors should be scrubbed only when the moon is waning. When the moon is waxing simply sweep the floor, or else mop it during a light sign. If you mop the floor during a water sign in Cancer, Scorpio, or Pisces, while the moon is waxing, moisture can get into the cracks, and eventually the wood may warp or even rot.

Cleaning Windows and Glass

Often streaks and smears are left behind after windows have been cleaned, even when ethyl alcohol is used. However, if you watch out for a light or warm day, when the moon is waning, water mixed with a dash of ethyl alcohol, rubbed off with newspaper will produce shining glass. Incidentally, when cleaning extremely dirty window frames you can achieve even better results on a water day. It is well worth the wait.

Porcelain

The Chinese, the discoverers of tea, have always taken it for granted that the dark coating that forms in the teapot should not be removed. They even say that the soul of the tea resides in this coating and every new cup adds a certain something, and this is what makes the brew 'a real pot of tea'.

Be that as it may, in many places, it is considered unrefined to place a delicate porcelain teapot, with a thick layer of tannin inside it on the tea table. This obstinate coating can be annoying, especially when attempts at cleaning it leave behind fine scratches on the precious porcelain, or the paint is removed. Those tormented with such things will take the following advice gratefully to heart. When the moon is waning, take a wet cloth, put a little salt on it and polish the stained surface of the porcelain.

When the moon is on the wane almost any gentle household remedy is helpful, whereas when the moon is waxing even aggressive scouring agents fail to produce the desired results and merely scratch the surface. The proof of the pudding is in the eating.

Metals

In many respects, the same rules apply to metals as to porcelain. On some days the polishing agent scratches patterns into the gleaming surface, while on other days it is enough to breathe on it and give it a wipe, and it's clean again.

Simply look out for the waning moon and use small quantities of mild household agents.

Metals should be cleaned when the moon is waning. For brass, stir equal parts of flour and salt with a little vinegar into a cream and apply thoroughly. Leave the cream for a short time to take effect, then wash it off and rub dry. Clean silver on an air day, with diluted ammonia water and polish afterwards with a little French chalk. For copper, stir a little salt into some hot vinegar and clean. Rub it dry afterwards.

SHOES

All sorts of footwear stay clean longer if they are polished when the moon is on the wane; the leather wears out less and lasts longer. Of course it's not possible only to clean shoes when the moon is waning, but obstinate dirt is easier to remove at that time. Particularly when winter boots are packed away in the cupboard, they should be cleaned and polished during the waning moon.

A first impregnation of brand new shoes when the moon is waning will last practically the whole life of the shoes.

MOULD

Modern, tightly shutting windows, combined with badly insulated walls have a serious disadvantage. The wall surfaces, especially in corners that let in the cold, can be a breeding ground for mould, if humidity is high. When the moon is waning, it is much easier to combat this than when it is waxing and the effects last longer. Mild remedies, such as vinegar and water, applied with a scrubbing brush, are quite sufficient.

SPRING CLEANING

Spring is the best breeding ground for a certain variety of bacillus. A strange restlessness seizes hold of the whole family. The lengthening days bring it to light: it's time for spring cleaning! Attic and dining-room, basement and garage are all waiting to be thoroughly ransacked, aired and cleaned.

Nature has so arranged things that the best days for such work happen to occur in spring (although there are other alternative dates that are almost as good throughout the year).

The ideal moment for clearing out, airing and thorough cleaning, is on an air day when the moon is waning. The air sign Aquarius that always falls in the waning moon, early in the year, is a particularly good time, because immediately afterwards the water sign Pisces aids all kinds of thorough cleaning.

Everything from Airing to Family Outings

VENTILATION

In this age of tightly shutting windows, rooms are not sufficiently ventilated, especially in winter. This is perhaps understandable, since one might well think that the air outside gives more cause for concern than the stale air indoors. Yet often precisely the reverse is the case. Nowadays people even talk of 'building sickness syndrome', a complex of illnesses that is caused by the poisonous vapours emanating from modern building materials, wood preservatives and air-conditioning units. Regular, natural ventilation is necessary and is always better than no ventilation at all.

The best time for thoroughly airing a building is on air and warmth days, while on earth and water days, a brief airing is sufficient.

Airing beds seems to have gone out of fashion, at least in the town, where these days it is a rare sight to see colourful duvets and feather beds billowing out from windows and balconies. There may be various reasons for this. Perhaps people are wary of allowing dust to blow into a neighbour's apartment on the floor below, or of damp weather getting into the feathers. For those prone to rheumatism, nothing could be worse.

Yet airing the bedding at the correct time is a good thing. The bedding becomes fresh and fragrant and lets the body breathe. Airing is also a good remedy against the dust mite, an allergenic microscopic animal that feeds on flakes of dead skin.

Bedding should be aired generously when the moon is on the wane, in an air or fire sign. Only air bedding briefly when the moon is waxing, otherwise too much moisture remains in the feathers. Avoid strong sunlight, as this can damage the feathers. When there is no 'r' in the month, the bedding can be aired for a longer time.

Mattresses are often kept in use for much too long (ten years or more). They should be regularly cleaned and above all aired. Airing them drives away the dust mites, which can only thrive in a moist, warm climate. Mattresses should be cleaned and aired when the moon is waning, preferably on an air day or a warmth day. This protects against vermin and draws out moisture (important for people with rheumatism or allergies). Under no circumstances should they be aired on a water day when the moon is waxing. This attracts damp and would be extremely bad for rheumatic people. Earth days are not especially suitable, either.

Storing away Summer and Winter Clothes

For those people who prefer not to use toxic preparations for keeping away moths, here is a well-tried natural remedy. Hang your summer or winter clothing away in the cupboard in autumn or spring, on an air day, when the moon is waning. Anti-moth preparations will then be superfluous. The ascending moon is a possible alternative. Clothing stored away on earth days can take on a rather strong smell and even go mouldy, while if it is stored away on a water day it will become damp.

Preserving, bottling, and storing

Nowadays home-made jam and home-preserved vegetables and fruit are all the rage. The key rules for successful harvesting, storing and preserving can be found in Chapter III.

A favourable time for making preserves is at the ascending moon, between Sagittarius and Gemini, when the fruit is much juicier and the aroma far better, too. Preserves will keep much longer and artificial setting agents will not be needed. Try this out once using your usual method and then on another occasion use less preserving sugar (say half the amount or you can use apples instead). The best time for gathering and preserving fruit is in Aries, on a fruit day. For root vegetables, the ideal moment is on root days in Capricorn and Taurus.

Fruit and vegetables keep better if they are frozen on a fruit day rather than on water days. They taste better when defrosted, do not become so watery and do not disintegrate.

BAKING

With pastries, cakes and biscuits the waxing moon on a light or fruit day brings better results. The dough remains manageable for longer. Baking wheat bread with yeast is most successful when the moon is waning on a flower day (Gemini, Libra and Aquarius). It will keep better too. When baking rye bread with sourdough the dough will not rise as easily when the moon is waning on a flower day. On the other hand, when it is waxing, the dough rises better and no extra leavening is needed.

Baking during Aries, Leo and Sagittarius produces better results. When baking, pre-heat the oven to a high temperature, and then lower the heat gradually. Try an experiment. Do one batch of baking when the moon is on the wane in Pisces, and another with the moon waxing in Leo. Then see the difference.

PAINTING AND VARNISHING

The demand for highly toxic and expensive paints has overtaken that for environmentally friendly, gentler lime wash paints and naturally produced varnishes. However, non-toxic paints are equally effective and durable particularly if they are applied at the right moment in the lunar cycle.

The period of the waning moon is the most suitable time for painting and varnishing. Over and undercoats dry well and form beautiful smooth surfaces. The top surface combines well with the undercoat and the work feels effortless.

Water days are unsuitable because the paint will not dry properly. Leo days are also unsuitable because the drying effect is too strong and sometimes causes the paint to crack.

HEATING UP THE HOUSE IN AUTUMN

Every autumn, the day finally comes when your own four walls need warming up, because the power of the sun is no longer sufficient to do this alone. In order to warm the whole house rapidly and thoroughly, the following rule should be observed. Turn the heating on for the first time in autumn on a warmth day in Aries, Leo or Sagittarius, when the moon is waning. This advice is particularly important for a new building, which should be heated without fail for the first time at the times specified above. This will drive the last dampness from the walls.

TRIPS TO THE COUNTRYSIDE AND THE FOUR DAY QUALITIES

Who does not enjoy an outing into the countryside now and then, whether alone, with a companion, or with family and friends?

Have you ever noticed that on day trips to the country, even when the outdoor temperature is the same, we 'feel' different – that we sometimes reach without thinking for our sunglasses even though the sky is cloudy, or that on one occasion we are happy to sit on the grass, while on another we daren't get out of our picnic chair because the ground feels unpleasantly damp or cold?

An example to illustrate this point occurs to me. What comes to mind is something that often happened to me as a boy years ago, when for many summer days in a row I used to go for rides on my first bicycle. With my pocket money I bought myself a drinking bottle – one of those plastic ones that you fix to the frame and can then drink from during the journey. I would be on the road with my friends for hours and was always surprised to find that with the same temperature conditions, the contents of the bottle would sometimes be all gone after an hour, while on other occasions I would bring the bottle home in the evening still half-full.

The solution to the puzzle is the four 'day-qualities'. These are particular characteristics of a day which are linked with the sign of the zodiac that is currently in force.

Warmth-days prevail when the moon is in the signs of Aries, Leo and Sagittarius. These are generally good days for an outing, and they feel warm even if the sky is cloudy. They have a drying effect, particularly in Leo, and on these days you will perhaps feel thirstier than usual. On Leo days, there often lurks the danger of strong storms suddenly brewing up, especially after long heat waves. Floods may follow as the earth is unable to take up enough moisture.

Light-days or air-days prevail when the moon is in Gemini, Libra or Aquarius. At this time the earth and plants absorb more light than usual and the effect on human beings is generally very pleasant. Car drivers occasionally find these days uncomfortable. Even when the sky is cloudy people may feel the need to wear sunglasses, as the light may seem piercing. Sportsmen, such as tennis players, sometimes find light days unpleasant if they have to serve into the sun, even if it is not shining directly into their eyes.

Cold-days or earth-days occur in Taurus, Virgo and Capricorn. On these days, even if the thermometer is registering quite high temperatures, one should always take slightly warmer clothes and blankets, in case the sun disappears behind the clouds. The earth feels cold to the touch and sometimes one will get goose-pimples even if the tiniest fleecy cloud passes in front of the sun. On these days you will probably come home with your drink's basket half-full.

Water-days occurring in Cancer, Scorpio and Pisces never allow the ground to dry out completely. On these days we are more likely to be in a hurry. If possible, don't leave home without waterproof clothing or an umbrella and maybe even a blanket, if you are planning to go on a picnic or to lie down on the ground after bathing.

Here's a tip. If you want to be better prepared for sudden changes in the weather or changes in climate, then watch out for the new moon and full moon and Gemini and Sagittarius days.

Care of the Body

Knowledge of the lunar cycles also can be applied to care of the body. The condition of a person's skin, hair and nails is an excellent indicator of their general state of health. Without 'beauty from within', above all, without a healthy diet, measures to take care of the body are frequently just cosmetic. They cover up the true causes of pale, fatty or blemished skin or of brittle nails.

The numerous tips and suggestions for a healthy way of life covered in Chapter II are a great help in working from within to produce healthy skin and strong hair, not least because you will then be able to save a lot of money on cosmetics which are often very expensive.

SKIN CARE

By skin care we do not mean the everday care of washing or using skin creams, but treatments for problematic skin conditions such as face packs or masks. Skin treatments are best carried out when the moon is waning. Minor operations to deal with bumps, pimples and so on should always be done at the waning moon as scars almost never form at this time. Skin specialists could avoid many problems if they would set flexible appointments during this period.

On the other hand, if you are applying substances to the skin, such as firming or moisturising creams, then the phase of the waxing moon is more suitable.

Anyone who wishes to take the sign of zodiac into account as well, ought not to miss Capricorn days. These are appropriate for every kind of skin care.

Hair Care

Many preparations for hair and the treatment of dandruff would be superfluous if you were to observe the correct times for hair treatment. At one time, no one was surprised if a barber kept his shop shut on certain days, because people knew that no one would be availing themselves of his services anyway. On the other hand, if Leo fell on a Sunday, many people would seek him out after church and entrust their hair to him.

In my home in Tyrol there used to be very few men with bald heads. Perhaps the reason for that was that from a baby's very first haircut, people always looked out for a Leo day.

The best moment for hair cuts are Leo and Virgo days, regardless of whether the moon is waning or waxing. On Pisces and Cancer days one should give haircuts a miss.

For anyone who is unhappy about his hair, because it is falling out, too thin or too greasy, we recommend this cure, which will not fail to be effective.

Have your hair cut or trimmed every month from February to August on Leo days. During that period Leo always occurs in the waxing moon, which further reinforces the beneficial effect. The Leo quality has a special effect on male hormones which may explain why hair is strengthened on Leo days. Once or twice a week during Leo, whisk one or two eggs (yolk and whites) and massage them into the hair after washing. Leave the mixture for a short while to take effect, then rinse it out with warm water, followed by a final rinse of cold water. Do not use a hair drier during this time. However, if this is absolutely necessary, wait for a quarter of an hour. Never blow-dry in the 'wrong' direction, or at too high a temperature as this damages the hair.

However, cutting hair exclusively on Leo days is not a guaranteed remedy for falling hair, because we often lose hair as a side-effect of medicines, due to hormonal changes and even stress. Hair loss also occurs frequently after pregnancy or during the menopause.

Hair that is cut during Virgo retains its shape and style for longer. Virgo is especially suitable for having a perm. During Leo the hair becomes too frizzy.

A haircut on Pisces days frequently leads to dandruff. If the hair is cut during Cancer it becomes shaggy and unmanageable.

If possible, you should also avoid washing your hair on Pisces and Cancer days. Many people, particularly young people, wash their hair practically every day; but in youth the body can put up with more. Later you may well wish to start taking this advice to heart.

If for any reason you should wish to say goodbye to bodily hair, then choose the time of the waning moon to remove it. For hair removal, the correct moment is in Capricorn, when the moon is on the wane (only during the first six months of the year). At that time the hair will not grow back so rapidly. However, you should not go too far with eyebrow-plucking.

CARE OF THE NAILS

If they are cut at the correct time, fingernails and toenails become hard, robust and do not break so easily. The most suitable time for cutting and filing nails is in Capricorn, or on any Friday after sunset.

The tip that you should care for your nails on Friday evenings will perhaps raise a smile at first; but my Grandfather, from whom I received this knowledge, used to swear by this rule and also said this was why he never had toothache. At any rate he lived to be 89 and throughout his whole life never had any problems with his teeth. I also followed this rule, together with my family, and can only confirm the connection. If you try it out, the good results will speak for themselves.

Guitarists are among the few people who have to attend to their nails very frequently. At the very least they should steer clear of unfavourable influences and dispense with nail-care on the first day of Pisces (directly after Aquarius) and on one of the two or three Gemini days.

In-growing nails should never be cut when the moon is waning, otherwise they always grow back incorrectly. The exception to this is correction of the nail-bed. This minor operation is more successful when carried out during the waning moon (avoiding Pisces if possible). Incidentally, the treatment of obstinate athlete's foot and fungal infections of the nails is also effective when the moon is on the wane. The same rule applies to warts as well.

MASSAGE

A skilled massage is not only soothing but also a very good preventive measure against illnesses of all kinds. It has a relaxing and stabilising effect on the heart and circulation, stimulates the activity of the organs and can be very helpful for people with blood pressure problems.

Where there is illness, special massages, such as lymph drainage can be particularly conducive to healing and decongestion. However, they should only be carried out by experienced physiotherapists. For massages that serve to relax, ease tension and detoxify, the most suitable time is when the moon is waning. Whereas if a massage is intended to have a mainly regenerating and strengthening effect, perhaps with the aid of appropriate oils, better results will be achieved when the moon is waxing.

In the course of conversations with my brother, Georg Koller, who has a physiotherapy practice in Osnabrück, Germany and who is also familiar with the lunar rhythms, I have learnt about the great success of special massages and how they are applied. As a physiotherapist and chiropractor he tries to carry out difficult treatments at the correct moment in the lunar cycle.

THE FEET

The feet are a very important and sensitive part of the body and are regrettably often neglected by many people. If our feet are not healthy, this manifests itself in our general well-being. Every area of the body is reflected in the reflex zones of the feet and can be influenced by reflexology massage. The rules for the ideal time to carry out this type of massage are the same as those for massage in general.

For the Future

All the rules in this book derive their validity from intuition and perception – not arbitrary whim, supposition, theory or faith. Sharpened senses, alertness and precise observation of nature and of their selves, made our ancestors 'masters of the art of timing'.

It would never have been possible to receive the knowledge of lunar rhythms and their influences from our ancestors and to pass it on again and again, if each succeeding generation had merely followed the rules without grasping their meaning. The ability of each successive generation to develop the awareness which confirms the validity of these rules and allows them to become flesh and blood (without always having to refer to a handbook or to call on the service of experts) has enabled this art to be passed on to us.

The position of the moon is merely the hand of a clock. The feeling for what it indicates is something that we bear within us. This book ultimately is only an aid to reawakening our awareness of nature's rhythms and regaining confidence in our innate knowledge of natural laws. This knowledge is valid and relevant everywhere on earth, but one has to grow with it organically. Our fields, like our bodies have had to adapt to so much negativity that the return to a natural way of life and harmony with the rhythms of nature demands time.

The lunar cycles can be of service to you at any time, if you familiarise yourself with a characteristic of nature, that she works slowly, at her own tempo and she will not let herself be hurried. If you keep this in mind, the knowledge of the lunar rhythms will come to you of its own accord.

This book is merely a tool, not a patent remedy. How you wield this tool is left entirely to you.

Appendix A

Herbs, plants and trees mentioned in the text

An asterisk indicates that the name refers to a number of related species.

English	Latin
Alder	Rhamnus frangula
Balm	Melissa officinalis
Basil	Ocimum basilicum
Birch	Betula★
Bramble	Rubus fruticosus
Chamomile	Matricaria recutita
Charlock	Sinapis arvensis
Chive	Allium schoenoprasum
Coltsfoot	Tussilago farfara
Comfrey	Symphytum officinale
Common Buckthorn	Rhamnus catharticus
Common Clubmoss	Lycopodium clavatum
Cornflower	Centaurea cyanus

Cowslip	Primula veris
Daisy	Bellis perennis
Dandelion	Taraxacum officinale
Dead-nettle	Lamium★
Elder	Sambucus nigra
Eyebright	Euphrasia officinalis
Fennel	Foeniculum vulgare
Fumitory	Fumaria officinalis
Garlic	Allium sativum
Greater Celandine	Chelidonium majus
Heartsease	Viola tricolor
Heather	Caluna vulgaris
Horsetail	Equisetum arvense
Lady's Mantle	Alchemilla★
Laurel	Laurus nobilis
Liverwort	Hepaticae★
Lovage	Levisticum officinale
Marigold	Calendula officinalis
Marjoram	Origanum vulgare
Mistletoe	Viscum album
Mugwort	Artemisia vulgaris
Mullein	Verbascum thapsus
Nasturtium	Tropaeolum majus
Parsley	Petroselinum crispum
Ribbed Melilot	Melilotus officinalis
Rosemary	Rosmarinus officinalis
Sage	Salvia officinalis

Senna	Cassia★
Shepherd's Purse	Capsella bursa-pastoris
Speedwell	Veronica officinalis
St John's Wort	Hypericum perforatum
Stinging Nettle	Urtica dioica
Sweet Chestnut	Castanea sativa
Sweet Woodruff	Asperula odorata
Tansy	Chrysanthemum vulgare
Thyme	Thymus vulgaris
Violet	Viola★
Watercress	Nasturtium officinale
Wild Radish	Raphanus raphanistrum
Willow	Salix★
Wood Garlic, Ramson	Allium ursinum
Wormwood	Artemisia absinthium
Yarrow	Achillea millefolium
Yellow Gentian	Gentiana lutea

Correspondence with the authors should be addressed to:

Johanna Paungger / Thomas Poppe
Post Box 107
A - 3400 Klosterneuburg
Austria
Email: tpoppe@compuserve.com
www.paungger-poppe.com

We shall endeavour to answer all enquiries, but due to the volume of letters we receive from all over the world, we cannot guarantee to reply to everyone. We are sincerely grateful for the confidence in our work that is expressed in these letters and would like to say a few words here in response.

There are many enquiries to which we are unable to reply for the simple reason that we do not know the answers. We are writing only from personal experience, and there are limits to that. This is true especially of physical and mental disorders. We are not doctors and we have no right – nor any wish – to presume to make judgements from a distance as to what is beneficial or harmful in a particular case.

We have received numerous requests for the addresses of good dowsers or healers who work according to lunar and natural cycles. Although the number of dowsers and healers is growing every day, all the ones that we know have now become hopelessly overstretched because their work is so successful. We can offer you some simple advice. If the doctor of your choice is unwilling to go along with your wishes, then find another one. A really good doctor will always do everything to ensure that you get well and stay well. On the other hand, anyone who works exclusively by the book and fixed rules, is either only interested in earning money, or else he is ignoring his own experience that statistics and patterns learnt by rote never cover the individual case.

A large proportion of the letters we received contained enquiries about sources for particular services or products in fields related to our work. With the aim of dealing with at least some of these requests, we looked for possible partners and companies over a period of time in order to help interested readers. Experience showed us that our aim could only be realised with the greatest difficulty. Firstly, many companies have become overstretched and booked up. They are no longer able to devote themselves to the individual customer in the necessary way. Secondly, in the course of time, the orientation of many of them has shifted away from service to the customer towards the greatest possible economic success. However, this rules out any close collaboration with us. Although there is nothing fundamentally wrong with money and economic success, what really matters is the way people handle these things. We take pleasure in success but never make it the goal of our work.

Many readers ask questions which have already been answered in the books, or which, with careful reading and patience, could be derived from the numerous basic rules. Our answer would only obstruct the pleasure and long-term profit to be gained from personal experience. The entire thrust of our second book is that our readers have to take the initiative and assume responsibility for their own health.

Most of the letters are ultimately about problems which the readers are asking us to help them solve. Yet almost always, the solution to the problem is already waiting just outside the door. Often the only reason why it is not allowed in is because in the search for it one has become obsessed by a certain direction and is then too proud, anxious or lazy to take a different tack.

Our entire work, both now and in the future is directed towards awakening in people the courage to make their own decisions and take responsibility for themselves – the courage to get right to the bottom of a problem, look at it from every side and think things through to the end. There is no other person, no expert who can take on this task for you – and that goes for us, too. If our work has been able to awaken in you the courage to do this, then we rejoice with you from the bottom of our hearts.

Index

Note: entries in **bold** refer to tables:
entries in *italic* refer to diagrams

Moon Calendar
2000–2010

♈	=	Aries
♉	=	Taurus
♊	=	Gemini
♋	=	Cancer
♌	=	Leo
♍	=	Virgo
♎	=	Libra
♏	=	Scorpio
♐	=	Sagittarius
♑	=	Capricorn
♒	=	Aquarius
♓	=	Pisces

☺	=	Full Moon
☾	=	Waning Moon
●	=	New Moon
☽	=	Waxing Moon

M	=	Monday
T	=	Tuesday
W	=	Wednesday
T	=	Thursday
F	=	Friday
S	=	Saturday
S	=	Sunday

2000

January	February	March	April	May	June
S 1	T 1	W 1	S 1	M 1	T 1
S 2	W 2	T 2	S 2	T 2	F 2 ●
M 3	T 3	F 3	M 3	W 3	S 3
T 4	F 4	S 4	T 4 ●	T 4 ●	S 4
W 5	S 5 ●	S 5	W 5	F 5	M 5
T 6 ●	S 6	M 6 ●	T 6	S 6	T 6
F 7	M 7	T 7	F 7	S 7	W 7
S 8	T 8	W 8	S 8	M 8	T 8
S 9	W 9	T 9	S 9	T 9	F 9 ☽
M 10	T 10	F 10	M 10	W 10 ☽	S 10
T 11	F 11	S 11	T 11 ☽	T 11	S 11
W 12	S 12	S 12	W 12	F 12	M 12
T 13	S 13 ☽	M 13 ☽	T 13	S 13	T 13
F 14 ☽	M 14	T 14	F 14	S 14	W 14
S 15	T 15	W 15	S 15	M 15	T 15
S 16	W 16	T 16	S 16	T 16	F 16 ☺
M 17	T 17	F 17	M 17	W 17	S 17
T 18	F 18	S 18	T 18 ☺	T 18 ☺	S 18
W 19	S 19 ☺	S 19	W 19	F 19	M 19
T 20	S 20	M 20 ☺	T 20	S 20	T 20
F 21 ☺	M 21	T 21	F 21	S 21	W 21
S 22	T 22	W 22	S 22	M 22	T 22
S 23	W 23	T 23	S 23	T 23	F 23
M 24	T 24	F 24	M 24	W 24	S 24
T 25	F 25	S 25	T 25	T 25	S 25 (
W 26	S 26	S 26	W 26 (	F 26 (	M 26
T 27	S 27 (	M 27	T 27	S 27	T 27
F 28 (	M 28	T 28 (	F 28	S 28	W 28
S 29	T 29	W 29	S 29	M 29	T 29
S 30		T 30	S 30	T 30	F 30
M 31		F 31		W 31	

2000

July	August	September	October	November	December
S 1 ●	T 1	F 1	S 1	W 1	F 1
S 2	W 2	S 2		T 2	S 2
	T 3	S 3	M 2	F 3	S 3
M 3	F 4		T 3	S 4)	M 4)
T 4	S 5	M 4	W 4	S 5	T 5
W 5	S 6	T 5)	T 5)		W 6
T 6		W 6	F 6	M 6	T 7
F 7	M 7)	T 7	S 7	T 7	F 8
S 8)	T 8	F 8	S 8	W 8	S 9
S 9	W 9	S 9		T 9	S 10
	T 10	S 10	M 9	F 10	
M 10	F 11		T 10	S 11 ☺	M 11 ☺
T 11	S 12	M 11	W 11	S 12	T 12
W 12	S 13	T 12	T 12		W 13
T 13		W 13 ☺	F 13 ☺	M 13	T 14
F 14	M 14	T 14	S 14	T 14	F 15
S 15	T 15 ☺	F 15	S 15	W 15	S 16
S 16 ☺	W 16	S 16		T 16	S 17
	T 17	S 17	M 16	F 17	
M 17	F 18		T 17	S 18 (	M 18 (
T 18	S 19	M 18	W 18	S 19	T 19
W 19	S 20	T 19	T 19		W 20
T 20		W 20	F 20 (	M 20	T 21
F 21	M 21	T 21 (	S 21	T 21	F 22
S 22	T 22 (	F 22	S 22	W 22	S 23
S 23	W 23	S 23		T 23	S 24
	T 24	S 24	M 23	F 24	
M 24 (	F 25		T 24	S 25	M 25 ●
T 25	S 26	M 25	W 25	S 26 ●	T 26
W 26	S 27	T 26	T 26		W 27
T 27		W 27 ●	F 27 ●	M 27	T 28
F 28	M 28	T 28	S 28	T 28	F 29
S 29	T 29 ●	F 29	S 29	W 29	S 30
S 30	W 30	S 30		T 30	S 31
	T 31		M 30		
M 31 ●			T 31		

2001

January	February	March	April	May	June
M 1	T 1 ☽	T 1	S 1 ☽	T 1	F 1
T 2 ☽	F 2	F 2		W 2	S 2
W 3	S 3	S 3 ☽	M 2	T 3	S 3
T 4	S 4	S 4	T 3	F 4	
F 5			W 4	S 5	M 4
S 6	M 5	M 5	T 5	S 6	T 5
S 7	T 6	T 6	F 6		W 6 ◯
	W 7	W 7	S 7	M 7 ◯	T 7
M 8	T 8 ◯	T 8	S 8 ◯	T 8	F 8
T 9 ◯	F 9	F 9 ◯		W 9	S 9
W 10	S 10	S 10	M 9	T 10	S 10
T 11	S 11	S 11	T 10	F 11	
F 12			W 11	S 12	M 11
S 13	M 12	M 12	T 12	S 13	T 12
S 14	T 13	T 13	F 13		W 13
	W 14	W 14	S 14	M 14	T 14 ☾
M 15	T 15 ☾	T 15	S 15 ☾	T 15 ☾	F 15
T 16 ☾	F 16	F 16 ☾		W 16	S 16
W 17	S 17	S 17	M 16	T 17	S 17
T 18	S 18	S 18	T 17	F 18	
F 19			W 18	S 19	M 18
S 20	M 19	M 19	T 19	S 20	T 19
S 21	T 20	T 20	F 20		W 20
	W 21	W 21	S 21	M 21	T 21 ◐
M 22	T 22	T 22	S 22	T 22	F 22
T 23	F 23 ◐	F 23		W 23 ◐	S 23
W 24 ◐	S 24	S 24	M 23 ◐	T 24	S 24
T 25	S 25	S 25 ◐	T 24	F 25	
F 26			W 25	S 26	M 25
S 27	M 26	M 26	T 26	S 27	T 26
S 28	T 27	T 27	F 27		W 27
	W 28	W 28	S 28	M 28	T 28 ☽
M 29		T 29	S 29	T 29 ☽	F 29
T 30		F 30		W 30	S 30
W 31		S 31	M 30 ☽	T 31	

2001

July	August	September	October	November	December
S 1	W 1	S 1	M 1	T 1 ☺	S 1
	T 2	S 2 ☺	T 2 ☺	F 2	S 2
M 2	F 3		W 3	S 3	
T 3	S 4 ☺	M 3	T 4	S 4	M 3
W 4	S 5	T 4	F 5		T 4
T 5 ☺		W 5	S 6	M 5	W 5
F 6	M 6	T 6	S 7	T 6	T 6
S 7	T 7	F 7		W 7	F 7 ☾
S 8	W 8	S 8	M 8	T 8 ☾	S 8
	T 9	S 9	T 9	F 9	S 9
M 9	F 10		W 10 ☾	S 10	
T 10	S 11	M 10 ☾	T 11	S 11	M 10
W 11	S 12 ☾	T 11	F 12		T 11
T 12		W 12	S 13	M 12	W 12
F 13 ☾	M 13	T 13	SS 14	T 13	T 13
S 14	T 14	F 14		W 14	F 14 ◉
S 15	W 15	S 15	M 15	T 15 ◉	S 15
	T 16	S 16	T 16 ◉	F 16	S 16
M 16	F 17		W 17	S 17	
T 17	S 18	M 17 ◉	T 18	S 18	M 17
W 18	S 19 ◉	T 18	F 19		T 18
T 19		W 19	S 20	M 19	W 19
F 20 ◉	M 20	T 20	S 21	T 20	T 20
S 21	T 21	F 21		W 21	F 21
S 22	W 22	S 22	M 22	T 22	S 22 ☽
	T 23	S 23	T 23	F 23 ☽	S 23
M 23	F 24		W 24 ☽	S 24	
T 24	S 25 ☽	M 24 ☽	T 25	S 25	M 24
W 25	S 26	T 25	F 26		T 25
T 26		W 26	S 27	M 26	W 26
F 27 ☽	M 27	T 27	S 28	T 27	T 27
S 28	T 28	F 28		W 28	F 28
S 29	W 29	S 29	M 29	T 29	S 29
	T 30	S 30	T 30	F 30 ☺	S 30 ☺
M 30	F 31		W 31		
T 31					M 31

2002

January	February	March	April	May	June
T 1	F 1	F 1	M 1	W 1	S 1
W 2	S 2	S 2	T 2	T 2	S 2
T 3	S 3	S 3	W 3	F 3	M 3 ☾
F 4			T 4 ☾	S 4 ☾	T 4
S 5	M 4 ☾	M 4	F 5	S 5	W 5
S 6 ☾	T 5	T 5	S 6		T 6
	W 6	W 6 ☾	S 7	M 6	F 7
M 7	T 7	T 7		T 7	S 8
T 8	F 8	F 8	M 8	W 8	S 9
W 9	S 9	S 9	T 9	T 9	
T 10	S 10	S 10	W 10	F 10	M 10
F 11			T 11	S 11	T 11 ●
S 12	M 11	M 11	F 12 ●	S 12 ●	W 12
S 13 ●	T 12 ●	T 12	S 13		T 13
	W 13	W 13	S 14	M 13	F 14
M 14	T 14	T 14 ●		T 14	S 15
T 15	F 15	F 15	M 15	W 15	S 16
W 16	S 16	S 16	T 16	T 16	
T 17	S 17	S 17	W 17	F 17	M 17
F 18			T 18	S 18	T 18 ☽
S 19	M 18	M 18	F 19	S 19 ☽	W 19
S 20	T 19	T 19	S 20 ☽		T 20
	W 20 ☽	W 20	S 21	M 20	F 21
M 21 ☽	T 21	T 21		T 21	S 22
T 22	F 22	F 22 ☽	M 22	W 22	S 23
W 23	S 23	S 23	T 23	T 23	
T 24	S 24	S 24	W 24	F 24	M 24 ☺
F 25			T 25	S 25	T 25
S 26	M 25	M 25	F 26	S 26 ☺	W 26
S 27	T 26	T 26	S 27 ☺		T 27
	W 27 ☺	W 27	S 28	M 27	F 28
M 28 ☺	T 28	T 28 ☺		T 28	S 29
T 29		F 29	M 29	W 29	S 30
W 30		S 30	T 30	T 30	
T 31		S 31		F 31	

2002

July	August	September	October	November	December
M 1	T 1 ☾	S 1	T 1	F 1	S 1
T 2 ☾	F 2	—	W 2	S 2	S 2
W 3	S 3	M 2	T 3	S 3	M 2
T 4	S 4	T 3	F 4	—	T 3
F 5	—	W 4	S 5	M 4 ●	W 4 ●
S 6	M 5	T 5	S 6 ●	T 5	T 5
S 7	T 6	F 6	—	W 6	F 6
—	W 7	S 7 ●	M 7	T 7	S 7
M 8	T 8 ●	S 8	T 8	F 8	S 8
T 9	F 9	—	W 9	S 9	—
W 10 ●	S 10	M 9	T 10	S 10	M 9
T 11	S 11	T 10	F 11	—	T 10
F 12	—	W 11	S 12	M 11 ☽	W 11 ☽
S 13	M 12	T 12	S 13 ☽	T 12	T 12
S 14	T 13	F 13 ☽	—	W 13	F 13
—	W 14	S 14	M 14	T 14	S 14
M 15	T 15 ☽	S 15	T 15	F 15	S 15
T 16	F 16	—	W 16	S 16	—
W 17 ☽	S 17	M 16	T 17	S 17	M 16
T 18	S 18	T 17	F 18	—	T 17
F 19	—	W 18	S 19	M 18	W 18
S 20	M 19	T 19	S 20	T 19	T 19 ☺
S 21	T 20	F 20	—	W 20 ☺	F 20
—	W 21	S 21 ☺	M 21 ☺	T 21	S 21
M 22	T 22 ☺	S 22	T 22	F 22	S 22
T 23	F 23	—	W 23	S 23	—
W 24 ☺	S 24	M 23	T 24	S 24	M 23
T 25	S 25	T 24	F 25	—	T 24
F 26	—	W 25	S 26	M 25	W 25
S 27	M 26	T 26	S 27	T 26	T 26
S 28	T 27	F 27	—	W 27 ☾	F 27 ☾
—	W 28	S 28	M 28	T 28	S 28
M 29	T 29	S 29 ☾	T 29 ☾	F 29	S 29
T 30	F 30	—	W 30	S 30	—
W 31	S 31 ☾	M 30	T 31		M 30
					T 31

143

2003

January	February	March	April	May	June
W 1 ♎	S 1 ♏ ●	S 1 ♏	T 1 ♐ ●	T 1 ♐ ●	S 1 ♊
T 2 ♐ ●	S 2 ♏	S 2 ♏	W 2 ♐	F 2 ♐	
F 3 ♐			T 3 ♐	S 3 ♐	M 2 ♋
S 4 ♐	M 3 ♒	M 3 ♒ ●	F 4 ♐	S 4 ♊	T 3 ♋
S 5 ♏	T 4 ♒	T 4 ♒	S 5 ♐		W 4 ♋
	W 5 ♒	W 5 ♐	S 6 ♊	M 5 ♊	T 5 ♌
M 6 ♏	T 6 ♐	T 6 ♐		T 6 ♋	F 6 ♐
T 7 ♒	F 7 ♐	F 7 ♐	M 7 ♊	W 7 ♋	S 7 ♌ ☽
W 8 ♒	S 8 ♐	S 8 ♐	T 8 ♊	T 8 ♋	S 8 ♌
T 9 ♐	S 9 ♐ ☽	S 9 ♐	W 9 ♋	F 9 ♐ ☽	
F 10 ♐ ☽			T 10 ♋ ☽	S 10 ♐	M 9 ♊
S 11 ♐	M 10 ♐	M 10 ♊	F 11 ♌	S 11 ♌	T 10 ♊
S 12 ♐	T 11 ♊	T 11 ♊ ☽	S 12 ♌		W 11 ♋
	W 12 ♊	W 12 ♊	S 13 ♐	M 12 ♌	T 12 ♋
M 13 ♐	T 13 ♋	T 13 ♋		T 13 ♊	F 13 ♎
T 14 ♊	F 14 ♋	F 14 ♋	M 14 ♌	W 14 ♊	S 14 ♎ ☺
W 15 ♊	S 15 ♋	S 15 ♌	T 15 ♌	T 15 ♋	S 15 ♐
T 16 ♊	S 16 ♐	S 16 ♐	W 16 ♊ ☺	F 16 ♋ ☺	
F 17 ♋			T 17 ♊	S 17 ♎	M 16 ♐
S 18 ♋ ☺	M 17 ♐ ☺	M 17 ♌	F 18 ♋	S 18 ♎	T 17 ♐
S 19 ♌	T 18 ♌	T 18 ♌ ☺	S 19 ♋		W 18 ♏
	W 19 ♌	W 19 ♊	S 20 ♎	M 19 ♐	T 19 ♏
M 20 ♌	T 20 ♊	T 20 ♊		T 20 ♐	F 20 ♒
T 21 ♌	F 21 ♊	F 21 ♋	M 21 ♎	W 21 ♏	S 21 ♒ ☾
W 22 ♌	S 22 ♋	S 22 ♋	T 22 ♐	T 22 ♏	S 22 ♐
T 23 ♊	S 23 ♋ ☾	S 23 ♎	W 23 ♐ ☾	F 23 ♒ ☾	
F 24 ♊			T 24 ♏	S 24 ♒	M 23 ♐
S 25 ♊ ☾	M 24 ♎	M 24 ♎	F 25 ♏	S 25 ♒	T 24 ♐
S 26 ♋	T 25 ♎	T 25 ♐ ☾	S 26 ♒		W 25 ♐
	W 26 ♐	W 26 ♐	S 27 ♒	M 26 ♐	T 26 ♐
M 27 ♋	T 27 ♐	T 27 ♐		T 27 ♐	F 27 ♊
T 28 ♎	F 28 ♏	F 28 ♏	M 28 ♒	W 28 ♐	S 28 ♊
W 29 ♎		S 29 ♏	T 29 ♐	T 29 ♐	S 29 ♊ ●
T 30 ♐		S 30 ♒	W 30 ♐	F 30 ♐	
F 31 ♐				S 31 ♊ ●	M 30 ♋
		M 31 ♒			

2003

July	August	September	October	November	December
T 1	F 1	M 1	W 1	S 1 ☽	M 1
W 2	S 2	T 2	T 2 ☽	S 2	T 2
T 3	S 3	W 3 ☽	F 3		W 3
F 4		T 4	S 4	M 3	T 4
S 5	M 4	F 5	S 5	T 4	F 5
S 6	T 5 ☽	S 6		W 5	S 6
	W 6	S 7	M 6	T 6	S 7
M 7 ☽	T 7		T 7	F 7	
T 8	F 8	M 8	W 8	S 8	M 8 🌕
W 9	S 9	T 9	T 9	S 9 🌕	T 9
T 10	S 10	W 10 🌕	F 10 🌕		W 10
F 11		T 11	S 11	M 10	T 11
S 12	M 11	F 12	S 12	T 11	F 12
S 13 🌕	T 12 🌕	S 13		W 12	S 13
	W 13	S 14	M 13	T 13	S 14
M 14	T 14		T 14	F 14	
T 15	F 15	M 15	W 15	S 15	M 15
W 16	S 16	T 16	T 16	S 16	T 16 ☾
T 17	S 17	W 17	F 17		W 17
F 18		T 18 ☾	S 18 ☾	M 17 ☾	T 18
S 19	M 18	F 19	S 19	T 18	F 19
S 20	T 19	S 20		W 19	S 20
	W 20 ☾	S 21	M 20	T 20	S 21
M 21 ☾	T 21		T 21	F 21	
T 22	F 22	M 22	W 22	S 22	M 22
W 23	S 23	T 23	T 23	S 23	T 23 🌑
T 24	S 24	W 24	F 24		W 24
F 25		T 25	S 25 🌑	M 24 🌑	T 25
S 26	M 25	F 26 🌑	S 26	T 25	F 26
S 27	T 26	S 27		W 26	S 27
	W 27 🌑	S 28	M 27	T 27	S 28
M 28	T 28		T 28	F 28	
T 29 🌑	F 29	M 29	W 29	S 29	M 29
W 30	S 30	T 30	T 30	S 30 ☽	T 30 ☽
T 31	S 31		F 31		W 31

2004

January	February	March	April	May	June
T 1	S 1	M 1	T 1	S 1	T 1
F 2		T 2	F 2	S 2	W 2
S 3	M 2	W 3	S 3		T 3 ☺
S 4	T 3	T 4	S 4	M 3	F 4
	W 4	F 5		T 4 ☺	S 5
M 5	T 5	S 6	M 5 ☺	W 5	S 6
T 6	F 6 ☺	S 7 ☺	T 6	T 6	
W 7 ☺	S 7		W 7	F 7	M 7
T 8	S 8	M 8	T 8	S 8	T 8
F 9		T 9	F 9	S 9	W 9 ☾
S 10	M 9	W 10	S 10		T 10
S 11	T 10	T 11	S 11	M 10	F 11
	W 11	F 12		T 11 ☾	S 12
M 12	T 12	S 13 ☾	M 12 ☾	W 12	S 13
T 13	F 13 ☾	S 14	T 13	T 13	
W 14	S 14		W 14	F 14	M 14
T 15 ☾	S 15	M 15	T 15	S 15	T 15
F 16		T 16	F 16	S 16	W 16
S 17	M 16	W 17	S 17		T 17 ☺
S 18	T 17	T 18	S 18	M 17	F 18
	W 18	F 19		T 18	S 19
M 19	T 19	S 20 ☺	M 19 ☺	W 19 ☺	S 20
T 20	F 20 ☺	S 21	T 20	T 20	
W 21 ☺	S 21		W 21	F 21	M 21
T 22	S 22	M 22	T 22	S 22	T 22
F 23		T 23	F 23	S 23	W 23
S 24	M 23	W 24	S 24		T 24
S 25	T 24	T 25	S 25	M 24	F 25 ☽
	W 25	F 26		T 25 ☽	S 26
M 26	T 26	S 27	M 26	W 26	S 27
T 27	F 27	S 28	T 27 ☽	T 27	
W 28	S 28 ☽		W 28	F 28	M 28
T 29 ☽	S 29	M 29 ☽	T 29	S 29	T 29
F 30		T 30	F 30	S 30	W 30
S 31		W 31		M 31	

146

2004

July	August	September	October	November	December
T 1	S 1	W 1	F 1	M 1	W 1
F 2 ☺	T 2	T 2	S 2	T 2	T 2
S 3	M 2	F 3	S 3	W 3	F 3
S 4	T 3	S 4	—	T 4	S 4
—	W 4	S 5	M 4	F 5 ☾	S 5 ☾
M 5	T 5	—	T 5	S 6	—
T 6	F 6	M 6 ☾	W 6 ☾	S 7	M 6
W 7	S 7 ☾	T 7	T 7	—	T 7
T 8	S 8	W 8	F 8	M 8	W 8
F 9 ☾	—	T 9	S 9	T 9	T 9
S 10	M 9	F 10	S 10	W 10	F 10
S 11	T 10	S 11	—	T 11	S 11
—	W 11	S 12	M 11	F 12 ☺	S 12 ☺
M 12	T 12	—	T 12	S 13	—
T 13	F 13	M 13	W 13	S 14	M 13
W 14	S 14	T 14 ☺	T 14 ☺	—	T 14
T 15	S 15	W 15	F 15	M 15	W 15
F 16	—	T 16	S 16	T 16	T 16
S 17 ☺	M 16 ☺	F 17	S 17	W 17	F 17
S 18	T 17	S 18	—	T 18	S 18 ☽
—	W 18	S 19	M 18	F 19 ☽	S 19
M 19	T 19	—	T 19	S 20	—
T 20	F 20	M 20	W 20 ☽	S 21	M 20
W 21	S 21	T 21 ☽	T 21	—	T 21
T 22	S 22	W 22	F 22	M 22	W 22
F 23	—	T 23	S 23	T 23	T 23
S 24	M 23 ☽	F 24	S 24	W 24	F 24
S 25 ☽	T 24	S 25	—	T 25	S 25
—	W 25	S 26	M 25	F 26 ☺	S 26 ☺
M 26	T 26	—	T 26	S 27	—
T 27	F 27	M 27	W 27	S 28	M 27
W 28	S 28	T 28 ☺	T 28 ☺	—	T 28
T 29	S 29	W 29	F 29	M 29	W 29
F 30	—	T 30	S 30	T 30	T 30
S 31 ☺	M 30 ☺		S 31		F 31
	T 31				

2005

January	February	March	April	May	June
S 1	T 1	T 1	F 1	S 1 ☾	W 1
S 2	W 2 ☾	W 2	S 2 ☾	M 2	T 2
	T 3	T 3 ☾	S 3	M 2	F 3
M 3 ☾	F 4	F 4		T 3	S 4
T 4	S 5	S 5	M 4	W 4	S 5
W 5	S 6	S 6	T 5	T 5	
T 6			W 6	F 6	M 6 ●
F 7	M 7	M 7	T 7	S 7	T 7
S 8	T 8 ●	T 8 ●	F 8 ●	S 8 ●	W 8
S 9	W 9	W 9	S 9		T 9
	T 10 ●	T 10 ●	S 10	M 9	F 10
M 10 ●	F 11	F 11		T 10	S 11
T 11	S 12	S 12	M 11	W 11	S 12
W 12	S 13	S 13	T 12	T 12	
T 13			W 13	F 13	M 13
F 14	M 14	M 14	T 14	S 14	T 14
S 15	T 15	T 15	F 15	S 15	W 15 ☽
S 16	W 16 ☽	W 16	S 16 ☽		T 16
	T 17	T 17 ☽	S 17	M 16 ☽	F 17
M 17 ☽	F 18	F 18		T 17	S 18
T 18	S 19	S 19	M 18	W 18	S 19
W 19	S 20	S 20	T 19	T 19	
T 20			W 20	F 20	M 20
F 21	M 21	M 21	T 21	S 21	T 21
S 22	T 22	T 22	F 22	S 22	W 22 ☺
S 23	W 23	W 23	S 23		T 23
	T 24 ☺	T 24	S 24 ☺	M 23 ☺	F 24
M 24	F 25	F 25 ☺		T 24	S 25
T 25 ☺	S 26	S 26	M 25	W 25	S 26
W 26	S 27	S 27	T 26	T 26	
T 27			W 27	F 27	M 27
F 28	M 28	M 28	T 28	S 28	T 28 ☾
S 29		T 29	F 29	S 29	W 29
S 30		W 30	S 30		T 30
		T 31		M 30 ☾	
M 31 ☾				T 31	

2005

July		August		September		October		November		December	
F 1		M 1		T 1		S 1		T 1		T 1	●
S 2		T 2		F 2		S 2		W 2	●	F 2	
S 3		W 3		S 3	●			T 3		S 3	
		T 4		S 4		M 3	●	F 4		S 4	
M 4		F 5	●			T 4		S 5			
T 5		S 6		M 5		W 5		S 6		M 5	
W 6	●	S 7		T 6		T 6				T 6	
T 7				W 7		F 7		M 7		W 7	
F 8		M 8		T 8		S 8		T 8		T 8	)
S 9		T 9		F 9		S 9		W 9	)	F 9	
S 10		W 10		S 10				T 10		S 10	
		T 11		S 11	)	M 10	)	F 11		S 11	
M 11		F 12				T 11		S 12			
T 12		S 13	)	M 12		W 12		S 13		M 12	
W 13		S 14		T 13		T 13				T 13	
T 14	)			W 14		F 14		M 14		W 14	
F 15		M 15		T 15		S 15		T 15		T 15	☺
S 16		T 16		F 16		S 16		W 16	☺	F 16	
S 17		W 17		S 17				T 17		S 17	
		T 18		S 18	☺	M 17	☺	F 18		S 18	
M 18		F 19	☺			T 18		S 19			
T 19		S 20		M 19		W 19		S 20		M 19	
W 20		S 21		T 20		T 20				T 20	
T 21	☺			W 21		F 21		M 21		W 21	
F 22		M 22		T 22		S 22		T 22		T 22	
S 23		T 23		F 23		S 23		W 23	(	F 23	(
S 24		W 24		S 24				T 24		S 24	
		T 25		S 25	(	M 24		F 25		S 25	
M 25		F 26	(			T 25	(	S 26			
T 26		S 27		M 26		W 26		S 27		M 26	
W 27		S 28		T 27		T 27				T 27	
T 28	(			W 28		F 28		M 28		W 28	
F 29		M 29		T 29		S 29		T 29		T 29	
S 30		T 30		F 30		S 30		W 30		F 30	
S 31		W 31								S 31	●
						M 31					

2006

January	February	March	April	May	June
S 1	W 1	W 1	S 1	M 1	T 1
M 2	T 2	T 2	S 2	T 2	F 2
T 3	F 3	F 3		W 3	S 3
W 4	S 4	S 4	M 3	T 4	S 4 ☽
T 5	S 5 ☽	S 5	T 4	F 5 ☽	
F 6 ☽			W 5 ☽	S 6	M 5
S 7	M 6	M 6 ☽	T 6	S 7	T 6
S 8	T 7	T 7	F 7		W 7
	W 8	W 8	S 8	M 8	T 8
M 9	T 9	T 9	S 9	T 9	F 9
T 10	F 10	F 10		W 10	S 10
W 11	S 11	S 11	M 10	T 11	S 11 ☺
T 12	S 12	S 12	T 11	F 12	
F 13			W 12	S 13 ☺	M 12
S 14 ☺	M 13 ☺	M 13	T 13 ☺	S 14	T 13
S 15	T 14	T 14	F 14		W 14
	W 15	W 15 ☺	S 15	M 15	T 15
M 16	T 16	T 16	S 16	T 16	F 16
T 17	F 17	F 17		W 17	S 17
W 18	S 18	S 18	M 17	T 18	S 18 ☾
T 19	S 19	S 19	T 18	F 19	
F 20			W 19	S 20 ☾	M 19
S 21	M 20	M 20	T 20	S 21	T 20
S 22 ☾	T 21 ☾	T 21	F 21 ☾		W 21
	W 22 ☾	W 22 ☾	S 22	M 22	T 22
M 23	T 23	T 23	S 23	T 23	F 23
T 24	F 24	F 24		W 24	S 24
W 25	S 25	S 25	M 24	T 25	S 25 ●
T 26	S 26	S 26	T 25	F 26 ●	
F 27			W 26	S 27	M 26
S 28	M 27	M 27	T 27 ●	S 28	T 27
S 29 ●	T 28 ●	T 28	F 28		W 28
		W 29 ●	S 29	M 29	T 29
M 30		T 30	S 30	T 30	F 30
T 31		F 31		W 31	

2006

July	August	September	October	November	December
S 1	T 1	F 1	S 1	W 1	F 1
S 2	W 2 ☽	S 2		T 2	S 2
	T 3	S 3	M 2	F 3	S 3
M 3 ☽	F 4		T 3	S 4	
T 4	S 5	M 4	W 4	S 5 ☺	M 4
W 5	S 6	T 5	T 5		T 5 ☺
T 6		W 6	F 6	M 6	W 6
F 7	M 7	T 7 ☺	S 7 ☺	T 7	T 7
S 8	T 8	F 8	S 8	W 8	F 8
S 9	W 9 ☺	S 9		T 9	S 9
	T 10	S 10	M 9	F 10	S 10
M 10	F 11		T 10	S 11	
T 11 ☺	S 12	M 11	W 11	S 12 ☾	M 11
W 12	S 13	T 12	T 12		T 12 ☾
T 13		W 13	F 13	M 13	W 13
F 14	M 14	T 14 ☾	S 14 ☾	T 14	T 14
S 15	T 15	F 15	S 15	W 15	F 15
S 16	W 16 ☾	S 16		T 16	S 16
	T 17	S 17	M 16	F 17	S 17
M 17 ☾	F 18		T 17	S 18	
T 18	S 19	M 18	W 18	S 19	M 18
W 19	S 20	T 19	T 19		T 19
T 20		W 20	F 20	M 20 ⊕	W 20 ⊕
F 21	M 21	T 21	S 21	T 21	T 21
S 22	T 22	F 22 ⊕	S 22 ⊕	W 22	F 22
S 23	W 23 ⊕	S 23		T 23	S 23
	T 24	S 24	M 23	F 24	S 24
M 24	F 25		T 24	S 25	
T 25 ⊕	S 26	M 25	W 25	S 26	M 25
W 26	S 27	T 26	T 26		T 26
T 27		W 27	F 27	M 27	W 27 ☽
F 28	M 28	T 28	S 28	T 28 ☽	T 28
S 29	T 29	F 29	S 29 ☽	W 29	F 29
S 30	W 30	S 30 ☽		T 30	S 30
	T 31 ☽		M 30		S 31
M 31			T 31		

2007

January		February		March		April		May		June	
M 1		T 1		T 1		S 1		T 1		F 1	☻
T 2		F 2	☻	F 2				W 2	☻	S 2	
W 3	☻	S 3		S 3		M 2	☻	T 3		S 3	
T 4		S 4		S 4	☻	T 3		F 4			
F 5						W 4		S 5		M 4	
S 6		M 5		M 5		T 5		S 6		T 5	
S 7		T 6		T 6		F 6				W 6	
		W 7		W 7		S 7		M 7		T 7	
M 8		T 8		T 8		S 8		T 8		F 8	☾
T 9		F 9		F 9				W 9		S 9	
W 10		S 10	☾	S 10		M 9		T 10	☾	S 10	
T 11	☾	S 11		S 11		T 10	☾	F 11			
F 12						W 11		S 12		M 11	
S 13		M 12		M 12	☾	T 12		S 13		T 12	
S 14		T 13		T 13		F 13				W 13	
		W 14		W 14		S 14		M 14		T 14	
M 15		T 15		T 15		S 15		T 15		F 15	◉
T 16		F 16		F 16				W 16	◉	S 16	
W 17		S 17	◉	S 17		M 16		T 17		S 17	
T 18		S 18		S 18		T 17	◉	F 18			
F 19	◉					W 18		S 19		M 18	
S 20		M 19		M 19	◉	T 19		S 20		T 19	
S 21		T 20		T 20		F 20				W 20	
		W 21		W 21		S 21		M 21		T 21	
M 22		T 22		T 22		S 22		T 22		F 22	☽
T 23		F 23		F 23				W 23	☽	S 23	
W 24		S 24	☽	S 24		M 23		T 24		S 24	
T 25		S 25		S 25	☽	T 24	☽	F 25			
F 26	☽					W 25		S 26		M 25	
S 27		M 26		M 26		T 26		S 27		T 26	
S 28		T 27		T 27		F 27				W 27	
		W 28		W 28		S 28		M 28		T 28	
M 29				T 29		S 29		T 29		F 29	
T 30				F 30				W 30		S 30	☻
W 31				S 31		M 30		T 31			

2007

July

- S 1
- M 2
- T 3
- W 4
- T 5
- F 6
- S 7 ☾
- S 8
- M 9
- T 10
- W 11
- T 12
- F 13
- S 14 ●
- S 15
- M 16
- T 17
- W 18
- T 19
- F 20
- S 21
- S 22 ☽
- M 23
- T 24
- W 25
- T 26
- F 27
- S 28
- S 29
- M 30 ☺
- T 31

August

- W 1
- T 2
- F 3
- S 4
- S 5 ☾
- M 6
- T 7
- W 8
- T 9
- F 10
- S 11
- S 12
- M 13 ●
- T 14
- W 15
- T 16
- F 17
- S 18
- S 19
- M 20
- T 21 ☽
- W 22
- T 23
- F 24
- S 25
- S 26
- M 27
- T 28 ☺
- W 29
- T 30
- F 31

September

- S 1
- S 2
- M 3
- T 4 ☾
- W 5
- T 6
- F 7
- S 8
- S 9
- M 10
- T 11 ●
- W 12
- T 13
- F 14
- S 15
- S 16
- M 17
- T 18
- W 19 ☽
- T 20
- F 21
- S 22
- S 23
- M 24
- T 25
- W 26 ☺
- T 27
- F 28
- S 29
- S 30

October

- M 1
- T 2
- W 3 ☾
- T 4
- F 5
- S 6
- S 7
- M 8
- T 9
- W 10
- T 11 ●
- F 12
- S 13
- S 14
- M 15
- T 16
- W 17
- T 18
- F 19 ☽
- S 20
- S 21
- M 22
- T 23
- W 24
- T 25
- F 26 ☺
- S 27
- S 28
- M 29
- T 30
- W 31

November

- T 1 ☾
- F 2
- S 3
- S 4
- M 5
- T 6
- W 7
- T 8
- F 9
- S 10 ●
- S 11
- M 12
- T 13
- W 14
- T 15
- F 16
- S 17 ☽
- S 18
- M 19
- T 20
- W 21
- T 22
- F 23
- S 24 ☺
- S 25
- M 26
- T 27
- W 28
- T 29
- F 30

December

- S 1 ☾
- S 2
- M 3
- T 4
- W 5
- T 6
- F 7
- S 8
- S 9 ●
- M 10
- T 11
- W 12
- T 13
- F 14
- S 15
- S 16
- M 17 ☽
- T 18
- W 19
- T 20
- F 21
- S 22
- S 23
- M 24 ☺
- T 25
- W 26
- T 27
- F 28
- S 29
- S 30
- M 31 ☾

2008

January	February	March	April	May	June
T 1	F 1	S 1	T 1	T 1	S 1
W 2	S 2	S 2	W 2	F 2	
T 3	S 3		T 3	S 3	M 2
F 4		M 3	F 4	S 4	T 3 ●
S 5	M 4	T 4	S 5		W 4
S 6	T 5	W 5	S 6 ●	M 5 ●	T 5
	W 6	T 6		T 6	F 6
M 7	T 7 ●	F 7 ●	M 7	W 7	S 7
T 8 ●	F 8	S 8	T 8	T 8	S 8
W 9	S 9	S 9	W 9	F 9	
T 10	S 10		T 10	S 10	M 9
F 11		M 10	F 11	S 11	T 10)
S 12	M 11	T 11	S 12)		W 11
S 13	T 12	W 12	S 13	M 12)	T 12
	W 13	T 13		T 13	F 13
M 14	T 14)	F 14)	M 14	W 14	S 14
T 15)	F 15	S 15	T 15	T 15	S 15
W 16	S 16	S 16	W 16	F 16	
T 17	S 17		T 17	S 17	M 16
F 18		M 17	F 18	S 18	T 17
S 19	M 18	T 18	S 19		W 18 ☺
S 20	T 19	W 19	S 20 ☺	M 19	T 19
	W 20	T 20		T 20 ☺	F 20
M 21	T 21 ☺	F 21 ☺	M 21	W 21	S 21
T 22 ☺	F 22	S 22	T 22	T 22	S 22
W 23	S 23	S 23	W 23	F 23	
T 24	S 24		T 24	S 24	M 23
F 25		M 24	F 25	S 25	T 24
S 26	M 25	T 25	S 26		W 25
S 27	T 26	W 26	S 27	M 26	T 26 (
	W 27	T 27		T 27	F 27
M 28	T 28	F 28	M 28 (	W 28 (	S 28
T 29	F 29 (	S 29 (	T 29	T 29	S 29
W 30 (		S 30	W 30	F 30	
T 31				S 31	M 30
		M 31			

2008

July	August	September	October	November	December
T 1	F 1 ●	M 1	W 1	S 1	M 1
W 2	S 2	T 2	T 2	S 2	T 2
T 3 ●	S 3	W 3	F 3		W 3
F 4		T 4	S 4	M 3	T 4
S 5	M 4	F 5	S 5	T 4	F 5 ☽
S 6	T 5	S 6		W 5	S 6
	W 6	S 7 ☽	M 6	T 6 ☽	S 7
M 7	T 7	M 8	T 7 ☽	F 7	
T 8	F 8 ☽	T 9	W 8	S 8	M 8
W 9	S 9	W 10	T 9	S 9	T 9
T 10 ☽	S 10	T 11	F 10		W 10
F 11		F 12	S 11	M 10	T 11
S 12	M 11	S 13	S 12	T 11	F 12 ☺
S 13	T 12	S 14		W 12	S 13
	W 13		M 13	T 13 ☺	S 14
M 14	T 14	M 15 ☺	T 14 ☺	F 14	
T 15	F 15	T 16	W 15	S 15	M 15
W 16	S 16 ☺	W 17	T 16	S 16	T 16
T 17	S 17	T 18	F 17		W 17
F 18 ☺		F 19	S 18	M 17	T 18
S 19	M 18	S 20	S 19	T 18	F 19 ☾
S 20	T 19	S 21		W 19 ☾	S 20
	W 20		M 20	T 20	S 21
M 21	T 21	M 22 ☾	T 21 ☾	F 21	
T 22	F 22	T 23	W 22	S 22	M 22
W 23	S 23	W 24	T 23	S 23	T 23
T 24	S 24 ☾	T 25	F 24		W 24
F 25 ☾		F 26	S 25	M 24	T 25
S 26	M 25	S 27	S 26	T 25	F 26
S 27	T 26	S 28		W 26	S 27 ●
	W 27		M 27	T 27 ●	S 28
M 28	T 28	M 29 ●	T 28	F 28	
T 29	F 29	T 30	W 29 ●	S 29	M 29
W 30	S 30 ●		T 30	S 30	T 30
T 31	S 31		F 31		W 31

2009

January	February	March	April	May	June
T 1	S 1	S 1	W 1	F 1 ☽	M 1
F 2			T 2 ☽	S 2	T 2
S 3	M 2	M 2	F 3	S 3	W 3
S 4 ☽	T 3 ☽	T 3	S 4		T 4
	W 4	W 4 ☽	S 5	M 4	F 5
M 5	T 5	T 5		T 5	S 6
T 6	F 6	F 6	M 6	W 6	S 7 ☺
W 7	S 7	S 7	T 7	T 7	
T 8	S 8	S 8	W 8	F 8	M 8
F 9			T 9 ☺	S 9 ☺	T 9
S 10	M 9 ☺	M 9	F 10	S 10	W 10
S 11 ☺	T 10	T 10	S 11		T 11
	W 11	W 11 ☺	S 12	M 11	F 12
M 12	T 12	T 12		T 12	S 13
T 13	F 13	F 13	M 13	W 13	S 14
W 14	S 14	S 14	T 14	T 14	
T 15	S 15	S 15	W 15	F 15	M 15 ☾
F 16			T 16	S 16	T 16
S 17	M 16 ☾	M 16	F 17 ☾	S 17 ☾	W 17
S 18 ☾	T 17	T 17	S 18		T 18
	W 18	W 18 ☾	S 19	M 18	F 19
M 19	T 19	T 19		T 19	S 20
T 20	F 20	F 20	M 20	W 20	S 21
W 21	S 21	S 21	T 21	T 21	
T 22	S 22	S 22	W 22	F 22	M 22 ☻
F 23			T 23	S 23	T 23
S 24	M 23	M 23	F 24	S 24 ☻	W 24
S 25	T 24	T 24	S 25 ☻		T 25
	W 25 ☻	W 25	S 26	M 25	F 26
M 26 ☻	T 26	T 26 ☻		T 26	S 27
T 27	F 27	F 27	M 27	W 27	S 28
W 28	S 28	S 28	T 28	T 28	
T 29		S 29	W 29	F 29	M 29 ☽
F 30			T 30	S 30	T 30
S 31		M 30		S 31 ☽	
		T 31			

2009

July	August	September	October	November	December
W 1	S 1	T 1	T 1	S 1	T 1
T 2	S 2	W 2	F 2		W 2 ☺
F 3		T 3	S 3	M 2 ☺	T 3
S 4	M 3	F 4 ☺	S 4 ☺	T 3	F 4
S 5	T 4	S 5		W 4	S 5
	W 5	S 6	M 5	T 5	S 6
M 6	T 6 ☺		T 6	F 6	
T 7 ☺	F 7	M 7	W 7	S 7	M 7
W 8	S 8	T 8	T 8	S 8	T 8
T 9	S 9	W 9	F 9		W 9 ☾
F 10		T 10	S 10	M 9 ☾	T 10
S 11	M 10	F 11	S 11 ☾	T 10	F 11
S 12	T 11	S 12 ☾		W 11	S 12
	W 12	S 13	M 12	T 12	S 13
M 13	T 13 ☾		T 13	F 13	
T 14	F 14	M 14	W 14	S 14	M 14
W 15 ☾	S 15	T 15	T 15	S 15	T 15
T 16	S 16	W 16	F 16		W 16 ●
F 17		T 17	S 17	M 16 ●	T 17
S 18	M 17	F 18 ●	S 18 ●	T 17	F 18
S 19	T 18	S 19		W 18	S 19
	W 19	S 20	M 19	T 19	S 20
M 20	T 20 ●		T 20	F 20	
T 21	F 21	M 21	W 21	S 21	M 21
W 22 ●	S 22	T 22	T 22	S 22	T 22
T 23	S 23	W 23	F 23		W 23
F 24		T 24	S 24	M 23	T 24 ☽
S 25	M 24	F 25	S 25	T 24 ☽	F 25
S 26	T 25	S 26 ☽		W 25	S 26
	W 26	S 27	M 26 ☽	T 26	S 27
M 27	T 27 ☽		T 27	F 27	
T 28 ☽	F 28	M 28	W 28	S 28	M 28
W 29	S 29	T 29	T 29	S 29	T 29
T 30	S 30	W 30	F 30		W 30
F 31			S 31	M 30	T 31 ☺
	M 31				

2010

January			February			March			April			May			June		
F	1	♑	M	1	♌	M	1	♌	T	1	♍	S	1	♎	T	1	♐
S	2	♑	T	2	♌	T	2	♌	F	2	♍	S	2	♎	W	2	♑
S	3	♌	W	3	♊	W	3	♊	S	3	♎				T	3	♑
			T	4	♊	T	4	♊	S	4	♎	M	3	♐	F	4	♒ ☾
M	4	♌	F	5	♍	F	5	♍				T	4	♐	S	5	♒
T	5	♌	S	6	♍ ☾	S	6	♍	M	5	♎	W	5	♑	S	6	♒
W	6	♌	S	7	♍	S	7	♎ ☾	T	6	♐ ☾	T	6	♑ ☾			
T	7	♊ ☾							W	7	♐	F	7	♑	M	7	♈
F	8	♊	M	8	♎	M	8	♎	T	8	♑	S	8	♒	T	8	♈
S	9	♍	T	9	♎	T	9	♐	F	9	♑	S	9	♒	W	9	♈
S	10	♍	W	10	♐	W	10	♐	S	10	♑				T	10	♈
			T	11	♐	T	11	♐	S	11	♒	M	10	♐	F	11	♊
M	11	♎	F	12	♑	F	12	♑				T	11	♐	S	12	♊ ●
T	12	♎	S	13	♑	S	13	♑	M	12	♒	W	12	♐	S	13	♊
W	13	♎	S	14	♑ ●	S	14	♒	T	13	♐	T	13	♐			
T	14	♐							W	14	♐ ●	F	14	♐ ●	M	14	♑
F	15	♐ ●	M	15	♒ ●	M	15	♒ ●	T	15	♐	S	15	♈	T	15	♑
S	16	♑	T	16	♒	T	16	♒	F	16	♐	S	16	♈	W	16	♌
S	17	♑	W	17	♒	W	17	♐	S	17	♐				T	17	♌
			T	18	♐	T	18	♐	S	18	♈	M	17	♑	F	18	♌
M	18	♑	F	19	♐	F	19	♐				T	18	♑	S	19	♌ ☽
T	19	♒	S	20	♐	S	20	♐	M	19	♈	W	19	♐	S	20	♊
W	20	♒	S	21	♐	S	21	♐	T	20	♑	T	20	♐			
T	21	♐							W	21	♑ ☽	F	21	♌ ☽	M	21	♊
F	22	♐	M	22	♈ ☽	M	22	♈	T	22	♐	S	22	♌	T	22	♍
S	23	♐ ☽	T	23	♈	T	23	♈ ☽	F	23	♐	S	23	♌	W	23	♍
S	24	♐	W	24	♑	W	24	♑	S	24	♌				T	24	♎
			T	25	♑	T	25	♑	S	25	♌	M	24	♊	F	25	♎
M	25	♐	F	26	♑	F	26	♐				T	25	♊	S	26	♎ ☺
T	26	♈	S	27	♐	S	27	♐	M	26	♊	W	26	♍	S	27	♐
W	27	♈	S	28	♐ ☺	S	28	♌	T	27	♊	T	27	♍			
T	28	♑							W	28	♍ ☺	F	28	♎ ☺	M	28	♐
F	29	♑				M	29	♌	T	29	♍	S	29	♎	T	29	♑
S	30	♌ ☺				T	30	♊ ☺	F	30	♍	S	30	♐	W	30	♑
S	31	♌				W	31	♊				M	31	♐			

158

2010

July	August	September	October	November	December
T 1 ♎	S 1 ♐	W 1 ♒ (	F 1 ♓ (	M 1 ♑	W 1 ♊
F 2 ♏		T 2 ♈	S 2 ♓	T 2 ♌	T 2 ♊
S 3 ♏	M 2 ♐	F 3 ♈	S 3 ♌	W 3 ♌	F 3 ♋
S 4 ♐ (	T 3 ♑ (	S 4 ♓		T 4 ♊	S 4 ♋
	W 4 ♐	S 5 ♓	M 4 ♌	F 5 ♊ ☺	S 5 ♍ ●
M 5 ♌	T 5 ♈		T 5 ♌	S 6 ♋ ☺	
T 6 ♐	F 6 ♈	M 6 ♌	W 6 ♌	S 7 ♋	M 6 ♍
W 7 ♑	S 7 ♓	T 7 ♌	T 7 ♊ ☺		T 7 ♈
T 8 ♐	S 8 ♓	W 8 ♌ ☺	F 8 ♊	M 8 ♍	W 8 ♈
F 9 ♈		T 9 ♌	S 9 ♋	T 9 ♍	T 9 ♐
S 10 ♈	M 9 ♌	F 10 ♊	S 10 ♋	W 10 ♐	F 10 ♎
S 11 ♋ ●	T 10 ♌ ●	S 11 ♊		T 11 ♐	S 11 ♎
	W 11 ♌	S 12 ♋	M 11 ♍	F 12 ♎	S 12 ♒
M 12 ♋	T 12 ♌		T 12 ♍	S 13 ♎)	
T 13 ♌	F 13 ♊	M 13 ♋	W 13 ♍	S 14 ♎	M 13 ♒)
W 14 ♌	S 14 ♊	T 14 ♍	T 14 ♐)		T 14 ♒
T 15 ♌	S 15 ♊	W 15 ♍)	F 15 ♐	M 15 ♒	W 15 ♈
F 16 ♌		T 16 ♐	S 16 ♎	T 16 ♒	T 16 ♐
S 17 ♋	M 16 ♋)	F 17 ♈	S 17 ♎	W 17 ♐	F 17 ♐
S 18 ♋)	T 17 ♋	S 18 ♐		T 18 ♌	S 18 ♐
	W 18 ♍	S 19 ♎	M 18 ♎	F 19 ♐	S 19 ♐
M 19 ♋	T 19 ♍		T 19 ♒	S 20 ♐	
T 20 ♋	F 20 ♐	M 20 ♎	W 20 ♒	S 21 ♐ ☺	M 20 ♈
W 21 ♍	S 21 ♐	T 21 ♒	T 21 ♐		T 21 ♈ ☺
T 22 ♍	S 22 ♐	W 22 ♒	F 22 ♐	M 22 ♈	W 22 ♋
F 23 ♍		T 23 ♒ ☺	S 23 ♐ ☺	T 23 ♈	T 23 ♋
S 24 ♐	M 23 ♎	F 24 ♈	S 24 ♐	W 24 ♈	F 24 ♌
S 25 ♐	T 24 ♎ ☺	S 25 ♈		T 25 ♓	S 25 ♌
	W 25 ♒	S 26 ♈	M 25 ♈	F 26 ♓	S 26 ♌
M 26 ♎ ☺	T 26 ♒		T 26 ♈	S 27 ♌	
T 27 ♎	F 27 ♒	M 27 ♈	W 27 ♈	S 28 ♌ (	M 27 ♌
W 28 ♎	S 28 ♐	T 28 ♈	T 28 ♓		T 28 ♊ (
T 29 ♒	S 29 ♐	W 29 ♈	F 29 ♓	M 29 ♌	W 29 ♊
F 30 ♒		T 30 ♈	S 30 ♓ (	T 30 ♌	T 30 ♋
S 31 ♐	M 30 ♈		S 31 ♌		F 31 ♋
	T 31 ♈